Cyber Outreach:

How to Develop and Evaluate Your Digital Ministries

**By
James R. Reed III and Lorrie C. Reed**

Cyber Outreach: How to Develop and Evaluate Your Digital Ministries

Copyright © 2020 by James R. Reed III and Lorrie C. Reed

No part of this book may be reproduced in any form or by any means without prior written permission of the authors or publisher, except as provided by United States of America copyright law.

ISBN: 978-1-7348375-2-0

All rights reserved

Printed in the United States of America

James and Lorrie Reed

Table of Contents

Preface iii
Chapter 1: Commissioned to Feed God's Sheep 1
Chapter 2: Digital Discipleship Defined 7
Chapter 3: Building Community 15
Chapter 4: Nourishing the People 26
Chapter 5: Planning for Success 40
Chapter 6: Steps in Strategic Planning 50
Chapter 7: Assessing and Evaluating the Plan 59
Chapter 8: Possibilities and Potential 67
Appendix A: 2014 Christian Education Survey Results 77
Appendix B: 2017 Christian Education Survey Results 79
Appendix C: Sample Course Evaluation 86
Appendix D: Assessing Environmental Support Factors 91
Appendix E: Sample Template for Social Media Metrics 93
Appendix F: Internet and YouTube Outreach by Program ... 94
Appendix G: Program Evaluation Questionnaire 95
About the Authors 100

Preface

There is one body and one Spirit, just as you were called in one hope of your calling; one Lord, one faith, one baptism; one God and Father of all, who is above all, and through all, and in you all. (Ephesians 4:4-6, NKJV).

We recall vividly the day authorities informed us we would have to "shelter in place." With the exception of essential workers, the rest of us were ordered to stay at home and carry on with life the best we could until the coronavirus passed over. We would be on lockdown until someone figured out when it would be safe for us to resume our routines. Among our prohibitions, we were not able to physically attend church services at our places of worship.

Now, several months into the pandemic, some churchgoers have stopped engaging in spiritual practices during this time of social distancing. Many of them are still longing for the day when they can get back to business as usual, however they choose to define that concept. The disconcerting part about all of this is that returning to the good old days may not be possible because this novel pandemic presents them with new challenges daily. Not being able to congregate as they did in pre-pandemic days, they are forced to find innovative ways to satisfy their need for spiritual nourishment and sacred community in a world that challenges them on many fronts. All of us find ourselves asking new questions:

- How do we redefine "church" in an era of physical isolation?
- How do we build a sacred community when we are distanced from each other?
- What is the purpose of Christian outreach?
- How do we care for each other spiritually while sheltering in place?

- How do we know if we are doing a good job?

At the same time, Christians continue to discover ways to satisfy their needs for the things that church has provided traditionally. According to a Barna Group study, "practicing Christians across the U.S. are seeking 'prayer and emotional support' (68% who have moved churches during [the pandemic], 52% who have stayed at their same church) and 'a Bible-centered message of hope and encouragement' (44% who have stayed at their same church, 35% all other practicing Christians) from their churches."[1]

Moreover, recent research shows that people who identify as Christian "agree strongly that faith is very important in their lives and attend church at least monthly [prior to the pandemic]—over half (53%) say they have streamed their regular church online within the past four weeks. Another 34 percent admits to streaming a different church service online other than their own, essentially 'church hopping' digitally."[2] Additionally, "Some congregations have gotten creative, not only with livestreaming services online or on television but by holding 'drive-in' religious services, where people can participate in services from cars that are spaced 6 feet apart, rather than inside a house of worship," according to a recent Pew research study. "To date, seven states explicitly allow this sort of gathering to take place, while many others encourage religious organizations to host services online."[3]

We conclude from all of this that many churchgoers who have stopped regularly attending worship services during the pandemic still desire "to have support from a church community."[4] At the same time, we submit to you that "business as usual" is a thing of the past. First, attendance and participation in institutionalized religion has been declining steadily for the past 10 years.[5] In our postmodern, multicultural society, the old norms have lost their relevance for a majority of young people. Therefore, a one-size-fits-all, "old time religion" no longer appeals to a significant segment of the church-going population. Third, after having been sheltered for so long, churches may have to continue to connect with people using innovative outreach programs. The audience for such programming is now worldwide due to the availability of popular cyber technology. Convenience is also a factor. We speculate that even after the pandemic is over and church doors are open once again, some people might prefer cyber outreach to brick-and-mortar worship or some hybrid combination of both.

The Purpose of this Book

In the previous paragraphs, we have identified some of the issues and concerns we attempt to address in our book, *Cyber Outreach: How to Develop and Evaluate Your Digital Ministries*. Written at a basic level, the book will serve as a "how to" text for anyone who is just starting to explore the use of technology for discipleship, worship, outreach, and other forms of service. The methods we describe have shown good success in small churches. All of the approaches we mention are simple and inexpensive to implement.

The overall purpose of this book is to provide guidance to small, cyber churches and other digital outreach initiatives that are trying to continue their ministries with limited financial resources. Our intended audience includes pastors, ministerial leaders, media coordinators, seminarians and lay people who are drawn to new ways of reaching out to audiences hungry for inspiration and hope by meeting the people where they are to be found.

Another prospective audience consists of people currently involved in discipleship who may want to explore new ways of "making disciples who make disciples." Those involved in digital outreach might use approaches described in this book to expand their current programs. Succinctly, this book will appeal to anyone who wishes to know more about spreading the gospel in the 21st century using the strengths and resources of the Internet.

Unique Contribution of this Book

Cyber Outreach: How to Develop and Evaluate Your Digital Ministries is a book that stands alone. No past or present text on the subject draws extensively on research, real-life examples, and organizational principles to raise awareness about the changing nature of Christian outreach in the present day. This resource provides practical explanations and definitions to help would-be leaders deploy cyberspace solutions and the social media to spread the gospel among people who may have shied away from church in recent years.

Organization of the Book

In our book, we will show you how to actualize important components that make for meaningful online communities. We also provide detailed guidance on how to evaluate your programs to ensure their effectiveness. Chapter 1 provides a rationale for increased interest in digital ministry. Chapter 2 describes popular social media platforms available for your use. Chapter 3 explains how we went about building community among the followers of our online chapel. Chapter 4 provides examples of how a small church expanded its outreach using the resources of the Internet. Chapter 5 explains the critical importance of having a mission, vision, goals and objectives that will serve as the umbrella under which everything else will fall. Chapter 6 shows you how to establish your online ministry using strategic planning as a model. Chapter 7 provides detailed examples of how to evaluate your program after it is established. Chapter 8 offers a model for program implementation using an action research approach. The book includes several helpful appendices.

Chapter 1: Commissioned to Feed God's Sheep

And Jesus came and spoke to them, saying, "All authority has been given to Me in heaven and on earth. Go therefore and make disciples of all the nations, baptizing them in the name of the Father and of the Son and of the Holy Spirit, teaching them to observe all things that I have commanded you; and lo, I am with you always, even to the end of the age." Amen. (Matthew 28:18-20, NKJV)

In a scripture known as the Great Commission, Jesus instructed his disciples to go into the world and make disciples of all nations (Matthew 28:16-20). Making disciples, both then and now, must involve nourishing the people in the myriad ways they need to be fed. In the 21st century, feeding God's people might involve preparing them to stand strong against the forces of oppression or racism or, equipping them to promote the kind of justice that Jesus taught. Nourishment might include a steady diet of encouragement for the least, the last, and the lost. Caring for God's people might further involve providing them with places where they can escape the madness going on all around them. Or it might involve simply praying for them. No matter how you fine tune it, God's people are hungry for spiritual food, and we, as 21st-century disciples and ministerial leaders, are responsible for feeding them. We are called to deliver the good news in novel ways, remaining starkly awake to the reality that one size no longer fits all (if it ever did!).

Altering styles and preferences for worship or expressing spirituality in different ways may be in order. Let's face it, given the declining attendance in brick-and-mortar mainstream churches, doing church might necessarily entail exploring options through the Internet as a way of influencing the broadest possible audience.

Decline of Institutionalized Religion

Given today's climate of declining support for organized religion, the task of making disciples can seem somewhat daunting. In 2015, the Pew Research Forum conducted a large-scale study of the Religious Landscape. According to the study of more than 35,000 U.S. adults, people who self-identify as atheists or agnostics, and people who believe in "nothing in particular"(a.k.a."nones") now make up roughly 23% of the U.S. adult population. The percentages of people who say they believe in God, pray daily and regularly go to church or other religious services all have declined modestly in recent years.[6] This trend is evident in a variety of demographic groups – across genders, generations and racial and ethnic designations.

Statistics and trend shifts, however, don't paint the whole picture. While the statistics cited here are, alarming, many people still view organized religion "as a force for good in American society."[7] Nearly nine out of ten adults say churches and other religious institutions bring people together and strengthen community bonds. Three out of four respondents say, "churches and other religious institutions help protect and strengthen morality in society."[8] Rainer reported recently, "Christians make up 2.3 billion people around the world. With numbers this large, the church isn't going away anytime soon."[9] The Internet holds a great deal of potential for not only sharing the Christian faith with those in need but also educating others about the Bible as a conceptual framework for righteous living.

Because the availability and use of the Internet is widespread, it makes sense to harness web technology to spread the Gospel message in non-traditional ways. In a recent Pew survey, members of congregations reported that they use the Internet "to strengthen the faith and spiritual growth of their members, evangelize and perform missions in their communities and around the world, and perform a wide variety of pious and practical activities for their congregations."[10]

The Pandemic of 2020 and Sheltering in Place

Another factor that has forced us to reach out in novel ways is the pandemic of 2020. It was to be a year to clarify our visions for the future and test a range of ministry possibilities for change and growth in the new decade. Many church leaders were excited and moved full steam ahead with spiritual energy. When the novel coronavirus forced everyone to slow down and reconsider our approach to tending to the spiritual needs of our congregations, many were unprepared to face the challenge.

No one had seen anything like it before. Those in the medical profession discovered new characteristics daily about this thing that had invaded our existence. After only a short time, the scientists and policy makers determined that the situation was so dangerous people would have to stay at home for their own safety and to shelter in place until the threat of danger had passed by. The whole idea of being on lockdown made many of us feel uneasy. Nevertheless, we managed to find new ways to serve the needs of God's people, some of whom had started to feel overburdened and fearful.

Whether online or on-ground, there is still much work to be done. So, we prayed a lot and opened our creative floodgates. We conferred with others internally and externally. Before long, our eyes were opened to fresh, new ways God was moving on behalf of God's people, even in the midst of a pandemic. The Prophet Isaiah once said of God, "I'm about to do a new thing... .Do you not perceive it?"(Isaiah 43:19)

The Body of Christ in the Digital Age

One of the first things many congregations accomplished was to reassure members that the body of Christ would not disappear. They were concerned there would no longer be a place of sanctuary to go to, a place of peace, a place of fellowship. Some were concerned that if they didn't physically attend church, they would be missing out on an opportunity to experience the Kingdom of God. Hence, our first job was to reassure them.

In the Gospel of Luke, Jesus told his disciples that the Kingdom of God is among us (Luke 17:21). It resides in the hearts and souls and minds of each and every believer. The Apostle Paul referred to the church as the body of Christ. He told the church in Ephesus that we are all one in the spirit. One body. One spirit. One faith. One Lord. One baptism. One God the Father of all, who is above all, through all, and in all and all (Ephesians 4:3). In other words, God is everywhere. And God is with us. And God is in us, even now as we shelter in place. So, while we were not able to worship in the brick-and-mortar building because we were ordered to maintain our social distancing, we reassured each other that the church would remain vibrant, and there would be no spiritual distancing among us. So, we proceeded to grow the church in the digital space.

Paradigm Shifts in the Church

Wilson discussed what it takes to grow a church. "During the late '40s and '50s, a growing church followed a growing population into the exploding suburbs. However, church growth was more of a sociological than a religious phenomenon. Going to church was what respectable people did, and associating with the right church was the way to get ahead."[11] In the '60s, things changed with the emergence of a secularized baby boom population, many of whom represented a "newly affluent middle class." Instead of attending church on Sundays, boomers opted for recreational activities. Canadian and American research studies showed that "most citizens still believed in God and still counted on the church for liturgical observances of life's transitions. They were not angry with the church, but felt no need to be part of a sustained Christian community."[12] Wilson noted that "in the '70s and '80s, at least in the Episcopal Church, we [the church] began to see some recovery." But the perceived recovery did not stem the decline. By the 2000s, millennials reported they had no belief in absolute standards for right and wrong,[13] as many of them continued to leave the church.

The Nature of Cyber Ministry

The 21st century cyber church is in a good position to spread the Gospel even in the digital space. Whether the ministry occurs in cyberspace or in person, this transformational vision can be accomplished in phases similar to those suggested by the Presbyterian Church (USA)[14]:

- First, the digital must to be a *place of sanctuary*, a place for seekers to find shelter and grace and work in conjunction with other community-based resource providers to ensure their spiritual needs are addressed.
- Second, the digital church must be an *empathetic community*, which focuses on the responsibility of the church to provide an environment in which seekers will be listened to without feeling shame or guilt.
- Third, the digital church must be a *covenantal community*, based on openness and trust, where people are bound together by a sense of common purpose.
- Finally, the digital church is called to be a *healing community*, a principle that emphasizes the physical, emotional and spiritual crisis associated with healing from the pandemic through a process of transformation.

These organizing principles can form the basis for a viable outreach ministry. Heidi Campbell has suggested that we should also pay attention to the needs of people we serve through the Internet. In a recent article, Campbell noted that over the past two decades, she has conducted "multiple studies looking at different manifestations of church online."[15] Through her interviews, she identified six traits that people value with regard to their online faith communities.[16] These are relationships, support and encouragement, sense of being appreciated, trusted connections, safe place for intimate communication, and shared beliefs and purpose. Other recent studies echo these traits, or ones similar to them.

In the next chapter, we will present information about digital discipleship. This will include a general definition of social media along with specific descriptions of some of the most popular digital platforms. We also describe the global context for social media use.

Chapter 2: Digital Discipleship Defined

Easton's Bible Dictionary defines the term *disciple* as - a scholar, sometimes applied to the followers of John the Baptist (Matthew 9:14) but principally to the followers of Christ. From at least one perspective, discipleship is the process of making better and more disciples. The Internet provides a mechanism to make this happen. According to Alexis, "the use of technology within our churches isn't new at all. The printing press is a technological tool which allowed God's word to spread throughout the world. Radio ads, print ads, flyers, mailers, etc. are all technological tools or products we used—and continue to use."[17]

When you really look at it, you can view the Internet as another tool for modern-day discipleship. Many congregants have stopped physically attending church but are still willing to watch a service online anonymously. One writer noted, "others have hearts that are ready to listen to spiritual messages but are intimidated by entering a church building." Using digital technology can extend the message to those who might find it otherwise inaccessible, thereby helping to spread the gospel to all nations.

Pew researchers revealed that people of faith found the Internet appealing for a number of reasons. Some of the reasons include its availability to provide answers to questions about faith issues, its ability to maintain anonymity of those making inquiries, its accessibility to provide information about religious organizations (staff, philosophy, activities, and so forth), its efficiency to link to other websites without having to self-generate content, and its capacity to facilitate communication for outreach to a much larger community.[18] And now, in the age of the pandemic, ministerial leaders can use the Internet to nourish people who are starving for spiritual sustenance in a world that no longer makes sense on many levels.

Using Social Media

Connecting with people on the World Wide Web comes in many forms.[19] One of those forms is social media. Social media is a term used to describe a collection of Internet-based communities that allow users to interact with each other online, according to Tech Terms. "This includes web-forums, wikis, and user-generated content websites." The term is most used to refer to the most popular of these sites, including: Facebook, YouTube, Twitter and Instagram.[20]

As shown in Figure 1, Facebook, Snapchat, Instagram, Twitter, and YouTube users visit these platforms frequently. Approximately 51% of Facebook users visit that platform several times a day. Following close behind is Snapchat with 48% of users reporting that they visit the platform several times a day. Users of other platforms report visiting several times a day; these include Instagram (38%), Twitter (26%), and YouTube (29%). So let's begin by defining which platforms are useful and giving an assessment of their advantages and limitations for cyber ministry.

A majority of Facebook, Snapchat and Instagram users visit these platforms on a daily basis

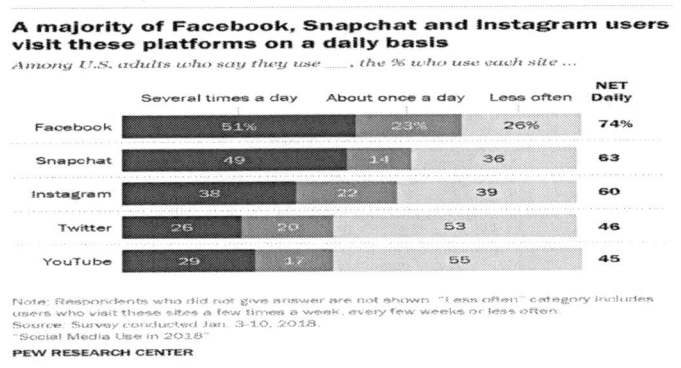

Among U.S. adults who say they use ___, the % who use each site ...

	Several times a day	About once a day	Less often	NET Daily
Facebook	51%	23%	26%	74%
Snapchat	49	14	36	63
Instagram	38	22	39	60
Twitter	26	20	53	46
YouTube	29	17	55	45

Note: Respondents who did not give answer are not shown. "Less often" category includes users who visit these sites a few times a week, every few weeks or less often.
Source: Survey conducted Jan. 3-10, 2018.
"Social Media Use in 2018"
PEW RESEARCH CENTER

Figure 1. Frequently visited social media platforms. Source: Pew Research Center.

Facebook

Tech Terms Computer Dictionary defines Facebook "as a social networking website that was originally designed for college students, but now is open to anyone 13 years of age or older. Facebook users can create and customize their own profiles with photos, videos, and information about themselves. Friends can browse the profiles of other friends and write messages on their pages." Sawaram Suthar, the head of marketing at Acquire, provides an excellent explanation of Facebook pros and cons.[21] He lists these in his article, "The Basic Advantage and Disadvantage of Facebook Ads."

Advantages:
- Facebook puts you in touch with millions of people from all walks of life.
- Facebook ad prices are cheaper than other forms of advertising.
- Facebook allows you to like and share material; the viral effect spreads the material.
- Facebook is organized so you can easily target people by age and/or location.
- Facebook enables you to like and share material widely for very little cost. There is the possibility of your material going viral.

Facebook Metrics

Facebook maintains certain metrics on the use of its platform. These metrics are called impressions, reach and engagement. Impressions are defined as the number of times your material is viewed. Reach is defined as the number of times your material is viewed by unique people. Finally, engagement is defined as how many times your material is liked, shared and/or receives comments. If consumers like and share the material to their friends, then that is called the viral effect.

Limitations:
- Can be difficult in catching people's attention – Suthar states, "People in Facebook are busy with their own stuff: communicating with friends, looking pictures and videos, and updating their own profile. Sometimes, ads are being snubbed by the busy users most especially if the ads are dull and plain. So if you are planning to put ads on Facebook, it requires a lot of creativity."
- It is very easy for people to post negative comments about your material, so you must be ready to handle the negative information.
- There is a lot of competition with other people's newsfeeds. This means you must focus on making sure your material stands out.
- Maintaining Facebook ads and pages requires a lot of time, resources, and energy.
- Facebook is in competition with other platforms.

Instagram

Tech Terms defines Instagram as "an online photo-sharing service. It allows you to apply different types of photo filters to your pictures with a single click, then share them with others." Instagram was designed to be predominately a mobile platform. In fact, all of the material posted to Instagram must occur on a smartphone or mobile device. The videos are limited to under one minute. Recently, Instagram added a video streaming channel called IGTV. However, there are limitations placed on consumers with respect to the orientation and length of their videos. Also, Instagram offers more privileges to verified users, but they determine which user is verified. Like Facebook, the rules are constantly decreasing one's ability to successfully manage these platforms. It has been widely reported that younger folks prefer Instagram and other predominately mobile platforms. Similarly to Facebook, Instagram uses the metrics of impressions, reach and engagements. Instagram was acquired by Facebook in 2012.

Twitter

Twitter is defined by Tech Terms as "an online service that allows you to share updates with other users by answering one simple question: "What are you doing?" In her article entitled "Twitter," Margaret Rouse defined Twitter as "a free social networking microblogging service that allows registered members to broadcast short posts called tweets."[22] Twitter members can broadcast tweets and follow other users' tweets by using multiple platforms and devices. Tweets and replies to tweets can be sent by cell phone text message, desktop client or by posting at the Twitter.com website. A shortcoming of the platform is that because Twitter limits each tweet to 140 characters, there is no room for rambling.[23] Hootsuite reports that Twitter users send over 500 million tweets per day. Twitter uses similar metrics as Facebook of impressions, reach and engagements.

Periscope

Periscope is a video streaming app that Twitter acquired in 2015. Elizabeth Muckensturm wrote in "The Pros and Cons of Twitter's Periscope App", "The social media realm develops so quickly that once we think we have it all down, something else surfaces."[24] The unique aspect of Periscope is that it is predominately a mobile platform. This reflects a growing trend among the consumers of social media in that the majority of their interactions occur over their smartphones, as compared to using laptops and/or desktop computers. As with Facebook, Periscope uses the metrics of impressions, reach, engagement as well as video views.

YouTube

Tech Terms defines YouTube as "a video sharing service that allows users to watch videos posted by other users and upload videos of their own. The service was started as an independent website in 2005 and was acquired by Google in 2006."

Podcasts

Tech Terms defines Podcast as "audio and video broadcasts that can be played on an iPod. However, because podcasts are downloaded using Apple iTunes and can be played directly within the program, you don't actually need an iPod to listen to a podcast."

Social Media Characteristics

Five years ago, social media was just getting started. Most people were not aware of and did not use social media. Email and paper memoranda were the dominant forms of business communication. Today, social media is used worldwide and has supplanted both emails and paper memoranda as the acceptable standard for business communication. Statisica reported substantial usage of social media worldwide.[25] Figure 2 presents the distribution by nation. In China, for example, in 2018, 673 million people used social media; the report projects 799.6 million users in 2023. In comparison, the United States, in 2018, had 243.6 million users with projections of 257.4 million users in 2023. Statista also reported that Facebook is the most used social media platform with more that 2 billion active users as of January 2019. YouTube is the second most used platform, with 1.9 billion users.

In this age where institutionalized Christianity is experiencing a declining presence[26] and the availability and use of the Internet is widespread, it makes sense to harness web technology to promulgate the Gospel message using social media and other non-traditional ways. Tapping into digital information networks increases the reach of the message. Fostering relationships online may stimulate some seekers to want to know more about the Good News they heard about on Facebook, Instagram or Twitter, to name a few.

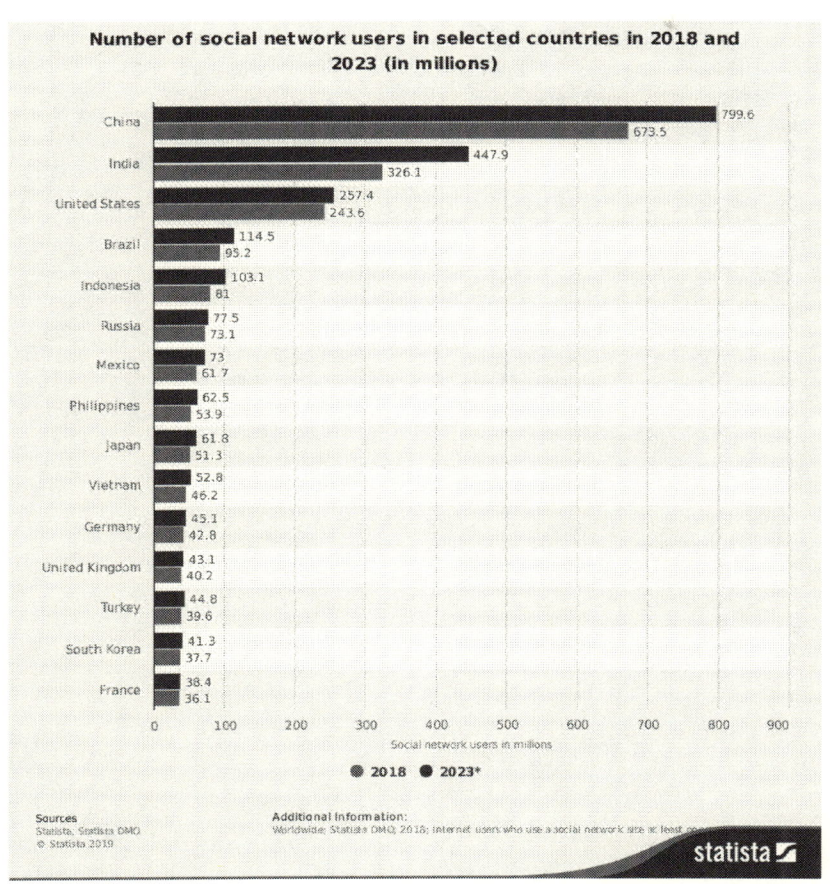

Figure 2. Statista projects trend of increasing usage worldwide. Source Statista.

Summary

The Internet holds great potential for cyber ministry at a time when institutionalized religion is declining. Social media, a collection of Internet-based communities, help users to connect and interact with each other online. Some of the most popular social media are Facebook, YouTube, WhatsApp, and

Facebook messenger. Each platform has its advantages as well as its disadvantages. For that reason, many users visit multiple platforms on a regular basis. Today, social media is used worldwide. It has replaced emails and other forms of paper messaging as an acceptable standard for communication.

The next chapter will describe how the paradigm is shifting with regard to building community online ministry. We describe our efforts to develop and grow a cyber outreach ministry. We provide a general history of the Rivertree Christian Chapel, including its mission and vision. We discuss how we grew the ministry by building community. The chapter includes tables depicting the results of our efforts.

Chapter 3: Building Community

The focus of this chapter is to describe steps we took to build community for our small, independent cyber ministry. Rivertree Christian Chapel was founded in 2014. The Chapel strives to serve God's people who are in need of a little peace in the busy and complex cyber world. The Chapel's ministries apply principles from the Bible as the basis for living in righteousness, wholeness, empowerment, critical reflection, and restoration. As a virtual chapel, Rivertree provides a quiet place to pray, reflect, and study.

The mission of Rivertree Christian Chapel was and is to glorify God and lift up the name of Jesus through sermons, prayer, discipleship, and education. Its vision is to establish a virtual sanctuary for prayer and reflection, maintain an online presence where people can seek shelter, disseminate inspirational messages to nourish people, and promote justice in the manner that Jesus taught his disciples

Built on a foundation of healing, renewal, and service, the Chapel endeavors to promote social change and carry out the Great Commission (Matthew 28:18-20). Through its programming, the Chapel offers several initiatives designed to "feed" God's people in the digital space. Among them are opportunities for prayer, daily lectionary, devotional readings, Christian education resources, and access to external self-care resources.

How the Christian Chapel Got Started

The Chapel started out as a private website for family members. We wanted to make sure our downline had a grounding in the Christian faith. So, we posted resources we thought would strengthen their faith and provide them with spiritual nourishment. At first, we included only the basics. The website included links to basic concepts like prayer, baptism, communion, the foundations of faith. Our hope was that the availability of this information would make a difference in the attitudes and behaviors of our children, grandchildren, great-grandchildren and other close relatives.

After the first year of operation, we began to see evidence that the social media message was getting through to our children and grandchildren based on the things they said, the times they relied upon their faith, and the times they incorporated Christian principles into their daily decision making. They shared with their friends. Before long, people outside of the immediate family started taking note of the Chapel. That's when we decided it was time to deliberately build our community.

In the beginning, we posted Bible verses, brief reflections, and posts from various sources. Honestly, there was little rhyme or reason for what we included on our site. We just knew we wanted to put something out there every day. Little by little, our following increased. After boosting the site to acquire additional members, our following increased to 745 people.

The Chapel now maintains a Facebook page and posts its offerings across other platforms, including Twitter, YouTube and Instagram. The Chapel keeps track of the rate of interaction among the users. We've seen people comment and become engaged in our daily posts. Visitors tend to share our material frequently with their friends and other connections. Plus, we find that there is reciprocity across the various platforms.

Building the Community

As our audience continued to grow, we became aware that we needed to provide something to fulfill the specific needs of our followers. We wanted our site to be more than a random collection of colorful posts, Bible verses, and humorous quips. We wanted our content to reflect our mission, vision, and values. So we deliberately changed our strategy. We formed a group and sent out invitations for people to like the page. We paid attention to the comments that were provided and to the activity from our followers in the form of likes and shares. We invited others to post content on the page as well.

Because of the algorithms Facebook uses, we noticed that our posts were not always reaching all of our followers. So, we decided to implement a strategy to establish a cohesive community in 2018. We started by paying attention to feedback in the form of likes, comments, and other engagements. Consequently, we decided to form a group, which eventually grew to the size of a moderately sized church. We tested out specific posts, videos, reflections, prayers, and graphics in an attempt to cater to the specific needs of our growing community. Along the way, we made adjustments in our content and included responses to situations that came up over time. Our offerings now include daily reflections, daily prayers, lectionary Bible readings, relaxing videos, external self-help resources for pastoral care, free educational materials and lesson downloads, sermons, a "thought for the day," and a feature prayer from a guest contributor.

Figure 3 below shows the Chapel's website Google metrics for the past three years. As shown in the graphic, between 2016 and 2019, the organic reach of the website was 272 people who viewed 3.61 pages per session. Over the three-year period these visitors have engaged in 6,828 sessions and have generated 24,648 page views. As of 2020, the Chapel's Facebook community group consisted of close to 750 people. The Chapel group consisted of close to 150 people. Currently, our main audience includes family members, professional colleagues, the sick and shut-in, the unchurched, those with waning interest in the brick-and-mortar church and active church members from a range of congregations.

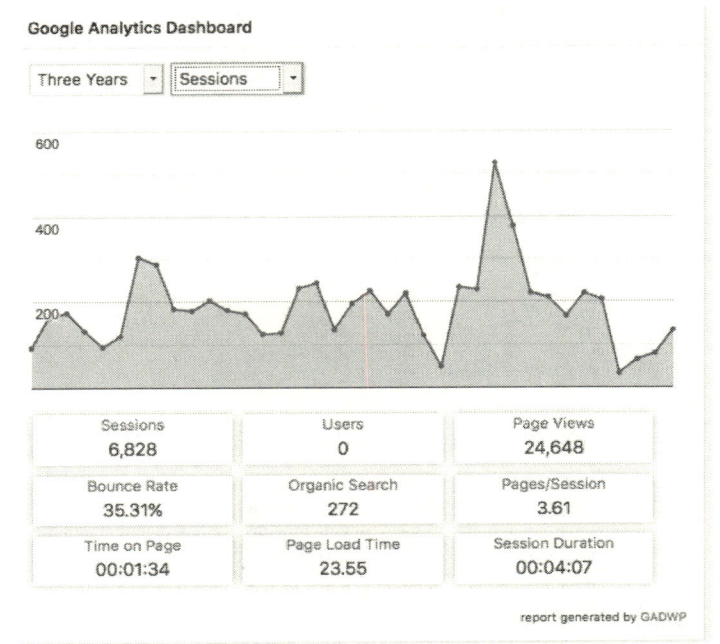

Figure 3. Google Analytics for Rivertree Christian Chapel website, 2016-2019.

Shut-in visitors are those who belong to a church, but for some physical or mental abnormality, they are unable to physically attend church services. Through several media platforms, we provide video recordings of sermons, praise songs, and Bible lessons. We also sent out emails with links to these platforms.

The unchurched visitors comprise a second target audience. People in this classification are no longer interested in organized religion. In some cases, they are people who choose not to affiliate with organized religion. For them, we provide social media posts, recordings, foundational information about the Christian faith, and positive daily affirmations depicted in a modern-day context. We reach members of this group primarily using various social media platforms, with special emphasis on the mobile platforms.

People with waning interest in the church represent a third target audience. These are people who may believe in Christ, but do not want to affiliate with an organized church for a range of reasons. For this group, we provide social media posts, recordings, and positive daily affirmations depicted in a modern-day context. As with the other groups we serve, we reach them using language and examples via various social media platforms, with special emphasis on the mobile platforms.

Finally, the Chapel appeals to active church members looking for enrichment resources for individual study, including daily prayer, several devotionals from different perspectives, and daily lectionary. People in this classification actively participate in ongoing church activities in their home congregational settings. They use the Chapel's resources to augment their spiritual nourishment.

We started to notice a trend. Our followers wanted daily prayers. They valued daily reflections about current affairs in matters of the soul. They appreciated the links to educational materials that we provided and many of them downloaded our free lesson plans. They liked our thoughts for the day, because they were presented in the form of videos with colorful and relevant content. We also began to include on our page posts from other groups that shared our values. For example, we began posting material from the United Church of Christ, Disciples of Christ, and groups that promote justice. We built our collection emphasizing content that focused on both Jesus and justice.

We also attempted to respond to meaningful events of the day. For example, during Black History Month, we provided a daily post highlighting a black history event that occurred during the month of February. Since 2020 was a leap year, we included 29 daily events and called the feature "29 Days." The response to those posts was very good.

In addition, we recognized the value of prayer. When the pandemic gripped the nation in February and March of 2020, we included a series of posts called "praying through the pandemic." Every day, we offered up a different prayer for those that were impacted.

We noticed that many of the people with whom we interacted were promoting either contemplative or deliverance prayer as the primary styles people should emphasize. However, based on the experience of people in our community, we knew that these styles did not always respond to the needs of the poor and disenfranchised people in our group. So, we formulated a devotional routine we called "praying from the bottom up." We published a journal by the same name and made it available on Amazon.com.

When protests over racial inequality and social injustice took center stage in the spring of 2020, we responded by providing basic educational pieces about the history of racism in America. These included links to reputable articles about slavery, reconstruction, Jim Crow, the great migration, the Civil Rights Movement, the criminal justice system, and other social concerns pertaining to the unequal treatment of Black people and other minorities in America.

Many of our social media posts originate from our website. On the website, we have included links to sermons Lorrie preached in other places. We also included a link to a 24/7 prayer line, along with links to external resources for pastoral care. As a result of our efforts, our community has grown, and its membership is comprised of people who share our values and promulgate the same.

Use of Facebook

Today's social media platforms allow for posts from every organization, large or small, to be viewed within each target's newsfeed. Therefore, the Chapel's posts and videos are competing with those from all sorts of ministries ranging from individuals with personal messages on up to mega churches with expensive digital resources. So, it's important that we emphasize high quality in our posts. We also make every effort to make sure our material is accurate and impactful on many levels. Because we have limited financial resources, we want to make sure that our investment is effective. Table 1 shows Facebook analytics for Rivertree Christian Chapel from January 1, 2018 to December 31, 2018.

Table 1. Rivertree Christian Chapel Facebook Metrics for 2018

Category	Posts	Reach	Impressions
Links	485	8154	14179
Photos	289	13850	24298
Shared Videos	43	4,366	6,315
Status	159	2,797	4,915
Videos	69	3,796	6,675
Unknown	6	541	830
2018 Totals	1,054	33,504	57,212

As shown in Table 1, the Chapel generated 1,054 posts in 2018. These posts reached 33,504 people. Reach means the number of unique individuals that have seen any page content. Individuals who were reached viewed the content 57,212 times. This number of displays is also called the number of impressions. The metrics in the table reflect different post types categorized by Facebook. Here are the definitions: Links are links to other people's posts that we shared on their page. Photos are photos that we posted on user's pages. Shared videos are videos that were posted by someone else and shared by us to others. Status is a figure reflecting the number of times we posted a written post (no photos or videos) provided to us from someone else. Videos are videos that we posted on users' pages.

We did a little research to discover the demographic of our followers. According to Facebook Insights, as of July 5, 2020, the audience for the Chapel consisted of 52% women and 47% men (refer to the Figure 4 below). In terms of geographical reach, out followers hail from the United States (n=699), Philippines (n=4), and other countries, including Australia, India, Guatemala, Ghana, and others (n=13+). We reach followers who reside in a number of U.S. cities, including Chicago, IL, New York City, Fort Worth, TX, Louisville, KY, Jacksonville, FL, Cleveland, OH, San Antonio, TX, Atlanta, GA, El Paso, TX, and Shreveport, LA.

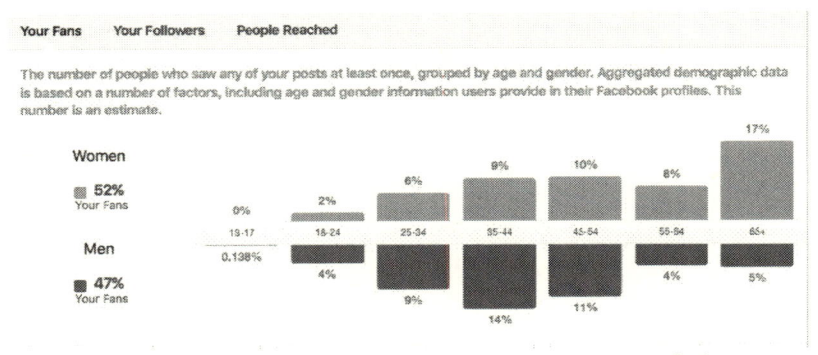

Figure 4. The number of people who saw any of the posts groped by age and gender.

Increasing Your Followers

As you can see, our followers grew in number in response to deliberate steps we took to promote expansion of our site. One sure-fire way to broaden your reach is to ask your followers to like and then share posts from your page. Having your post in their newsfeed increases the likelihood that they will see it, read it, and respond to it in some way.

Another way to expand your reach on Facebook is to boost your posts. Boosting is easy. First you determine your goals; in other words ask, what is your reason for getting your post into the hands of more readers? You may decide that your goal is to have your posts seen on a regular basis, which would suggest that you should boost your posts through automated ads. This option is used frequently for businesses. Other reasons you may want to boost your post would include your desire to (a) get more leads, (b) get more website visitors, (c) promote your page, or (d) draw more attention to a particular post.

When you promote your page or boost a particular post, you will want to target your distribution. Facebook gives you the ability to select a particular geographical location for your ad, identify a specific age range, determine your daily budget for your ad, and set a duration for the ad to run. Facebook will automatically bill you at the end of your campaign.

Teach One Inspirations

Teach One Inspirations is another source of posts for the Chapel's Facebook page. Jim created Teach One as a way to provide positive influences that can shared with others from a Christian perspective. He started creating his graphics as a hobby. Recently, however, generating these messages has grown into a personal ministry for Jim. Table 2 shows Teach One Inspiration metrics for 2018. Of the 661 posts from Teach One, we reached 37,490 people who opened the content 60,396 times.

Table 2. Teach One Inspirations Facebook Metrics for 2018

Teach One Influences	Posts	Reach	Impressions
Links	73	2,381	3,459
Photos	393	17,004	28,039
Shared Videos	30	116	230
Status	5	16	42
Videos	160	17,973	28,626
2018 Totals	661	37,490	60,396

Summary

This chapter described how the Rivertree Christian Chapel built a viable community focused on providing a safe and quiet place for people to rest up from the daily "noise" of the Internet. The Chapel represents a positive influence with the potential for canceling out negative influences that visitors may have encountered elsewhere. The Chapel in cyberspace seeks to feed God's sheep by providing materials for daily prayer, Bible reading, and devotional reading. The website also contains links to external resources visitors might use to fulfill other needs in their lives.

The chapel started out as a website for family members with the goal of increasing the likelihood they would have access to accurate information about their faith. Soon, the

Cyber Outreach

website and Facebook page began to attract people outside of the immediate family. Currently, our reach is international in scope.

The next chapter will describe the outreach efforts of a brick-and-mortar church seeking to expand it cyber ministries. In the chapter, we begin by providing background on the Grace United Church of Christ. We share the process they used to move toward digital outreach. We describe in detail their efforts to continue outreach in the face of a pandemic that resulted in prohibition of large gatherings. Finally, we provide tips on how to avoid some of the technical pitfalls they encountered.

Chapter 4: Nourishing the People

Background for Grace United Church of Christ

Grace United Church of Christ is a new church start planted by Covenant United Church of Christ and nurtured by the Illinois Conference and the Eastern Association. The church opened its doors for the first time in 2014. The Pastor, the Rev. Melody L. Seaton, stated that the church's desire was to serve God and God's people by spreading the Gospel message through the preached word, Bible study, and a music ministry; provide activities and services that enrich the body, mind and soul; create an environment for cultivating leaders for ministry and service; and provide a safe haven for children of all ages to grow and prosper in the ways of the Lord. The congregation strives to be "witnesses to the Gospel of Jesus Christ," and endeavors for truth, justice and peace in our community, nation and world. Early in 2015, the church wrote and ratified a constitution. The document described the institution as a "self-governing congregation of seekers guided by the Scriptures" and by the denomination's Statement of Faith.

Values: "I See People"

In one of her reflections, the Pastor stated her personal vision was "I see people." Accordingly, she sponsored events to attract people to the church. For example, the church hosted lavish dinners and invited friends and family to fellowship services. The church's programming included things like Jammin' for Jesus (a back-to-school community fair), "Share the Warmth" (distributing outerwear and accessories to children in the community regardless of their membership status), and a prom dress giveaway. Additionally, the Pastor often participated in community prayers in conjunction with other churches in the municipality. She also established a food pantry to benefit low-income families regardless of their membership status.

In spite of all these efforts, the people still did not fill the pews on any given Sunday morning. Nor did these efforts result in increased membership. When people were not joining or attending at the rate she expected, the Pastor decided it was time for a change in strategy.

Evolution toward Digital Outreach

One of the first steps the Pastor took as part of her new strategy was to hire a part-time media/technology coordinator. Serving in this role at Grace UCC, co-author Jim believes his work with the social media marketing was a calling. He said, "I have seen growth in the use of the social media over time. We didn't start out this way five years ago. As a matter of fact, when we first started, I had no idea where all this was headed. I was just simply trying to advertise for the church because I was asked to do so. Over time, the posts have improved. If you look at the posts that I did five years ago as compared to today, there is a significant difference." Jim also noted that he was beginning to see that social media was becoming important as a tool for discipleship.

Major Initiatives

In a focused approach, the church emphasized three major initiatives; namely, preaching, singing, and teaching to the glory of God.

Preaching

As a former educator, the Pastor served not only as a spiritual leader but also as a teacher and mentor to aspiring ministerial leaders in her congregation. Over the years of the church's existence, she has nurtured interns, seminary students, deacons, youth leaders, even associate pastors. Each person has brought his or her own style and voice to the pulpit. According to one media expert, "like attracts like." Hence, the variety of speakers appealed to different audiences, and the sermons reached a broader demographic. The variety further increased the likelihood that the church would receive notice from people outside the brick-and-mortar church.

The Pastor encouraged the media coordinator to use metrics available through social media to track the number of people who followed the sermons. She placed a priority on maintaining equipment to ensure that sermon broadcasts were of high quality. This approach included replacing faulty pulpit microphones and purchasing additional wireless microphones. She realized that social media provided for real time broadcasts as well as rendering historical evidence of viewership. She reported on these numbers in her quarterly meetings. Congregants were able to not only see numbers, but various categories of viewership along with geography. Given these metrics, the media coordinator was able to make the necessary adjustments in what the church provided and who the target audience was.

Music

In terms of music, Grace UCC attracted a great deal of Internet attention from weekly videos of the church's Praise Team. Every Sunday, this dynamic group of musicians set the tone for worship. In addition to vocalists, the church's instrumental musicians are among the best around! Some days the Pastor jokingly said that there was probably no need for an additional sermon because the Praise Team had already delivered the message. The media coordinator recorded their worship songs, which were broadcast on social media and distributed in the church's weekly email.

Christian Education

In terms of education, the Pastor believed that discussing the Bible in community contributes to a rich understanding of how the scripture has affected us all. By sharing each other's perspectives, congregation members all grow and learn together. The Christian Education Department used this approach in Sunday school as well as Thursday Night Bible study (see Table 3). The mid-week Bible classes were broadcast via Live Stream on Facebook. Videos of the classes also were available for viewing by the public on demand through the church's archives and YouTube.

Table 3. Thursday Night Bible Study Facebook Metrics

Year	Reach	Impressions
2015	14,224	30,121
2016	160,120	275,878
2017	130,871	242,161
2018	173,941	195,546

Importance of Social Media

Looking for church growth in the digital space involved use of major digital platforms for video broadcasts of sermons and the Praise Team. The church also used these and other platforms to broadcast "real-time" Thursday Night Bible Study classes via Live Stream (Facebook, YouTube, Periscope, Website). Weekly broadcasts were also included in the church's newsletters and distributed through emails.

Table 4 summarizes social media metrics for 2018 by area of emphasis. The first column shows the number of videos posted for each category; that is, Praise Team, Bible study, and sermons. Data are reported in terms of posts, reach, and impressions. Reach represents the number of people who received the content in their news feed. Impressions shows the number of times those videos were viewed. The media coordinator monitored the data continually and adjusted techniques based on observed feedback from the target audience. The media coordinator created both social media and email campaigns to reach a targeted audience on a cost-effective basis.

Table 4. Social Media Metrics for Grace in 2018 by Area of Emphasis

Category	Video Posts	Reach	Impressions
Praise Team	52	7,689	12,767
Bible Study	61	6,396	12,174
Sermons	66	17,766	34,743

Table 5 shows overall Facebook metrics for 2015-2018. As shown in the table, Grace reached 479, 156 people during the 4-year period. Reach means the number of unique individuals that have seen any page content. Individuals who were reached viewed the content 743,706 times. This number is also called the number of impressions.

Table 5. Overall Facebook metrics 2015- 2018

Year	Reach	Impressions
2015	14,224	30,121
2016	160,120	275,878
2017	130,871	242,161
2018	173,941	195,546
4-Year Total	479, 156	743,706

Responding to the Pandemic

As with most congregations during the first 6 months of 2020, Grace UCC was temporarily closed due to the state's shelter-in-place order. Public health officials banned gatherings of more than ten people, required the wearing of face masks, and recommended social distancing of at least six feet between congregants. Consequently, during the coronavirus pandemic, our houses of worship became places where we could no longer gather together.

Cyber Outreach

Grace UCC adjusted its outreach efforts in various ways. During this time, the Pastor maintained virtual office visits. Congregants were able to contact her by email, cellphone or by leaving a message on the church's phone; she later contacted them via phone, text, Facetime, Zoom, or Google Hangout.

Additionally, the weekly schedule changed. Some of those changes are described in the next few pages.

Prayer Call

The church instituted a Sunday morning prayer line, which began at 9:30 a.m. Members received a telephone number and access code for dialing into the line. The call lasted approximately one hour.

Leaders of other congregations have frequently asked the Pastor how she managed to hold her congregation together when so many other churches had become fragmented. The weekly prayer call was one of those initiatives. The Pastor recruited members of the congregation the week prior to each call to ask them to read a scripture or say a prayer during the call. The agenda typically included updates on current events, scripture reading and prayer, prayer requests and praise reports. Involving church member in this way contributed to the congregation's sense of community, promoted spiritual unity, and helped congregants develop skills for ministerial leadership.

Cyber Church via Livestream on Facebook

Livestream broadcasts via Facebook replaced the face-to-face, brick-and-mortar experience congregation members had enjoyed in pre-pandemic days. By Friday of each week, appointed members of the congregation sent the media coordinator prerecorded worship segments to be included in the week's broadcast. A welcome by a young person or other church member, a call to worship, offering, altar prayer, sermon, and benediction comprised the service. The media coordinator, using resources available through Final Cut Pro, spliced the pieces together, inserted transitions, rendered the recording, and scheduled it for broadcast. The service typically appeared on several platforms simultaneously, including Facebook, YouTube, Twitter, and the church's website. The coordinator monitored and responded to comments from viewers. Prior to the pandemic, the broadcast reached approximately 60 people weekly. However, during the pandemic, the Sunday Morning Worship broadcast typically reached more than 150 people every week.

Thursday Night Bible Study via Zoom

The church continued to offer Thursday Night Bible Study at 7:00 p.m. on Thursdays. At first, the class consisted of rebroadcasts of pre-pandemic lessons. However, during their time of sheltering, the church began to conduct the class live via Zoom. Each week, a member was appointed to facilitate the class, ensuring that all learners had opportunities to participate. On average, around eight people attended the Zoom sessions, which were subsequently viewed by at least 75-100 people via the Internet each week.

Cyber Outreach

Public Service Webinars

On Wednesdays at 7:30 p.m., the Pastor hosted a series of Facebook webinars entitled, "Faith & Fear." Guest speakers included experts on topics that were of interest to members of the congregation. These sessions attracted approximately 100 people on average.

"Faith & Fear" - "Suffer the Little Children" (106 Views)

The emphasis of the "Suffer the Little Children" session was on how to help children navigate the numerous pandemics playing out before their eyes. The presenter was Dr. Matowe from the Family Institute of Northwestern University in Chicago. The session used a question-answer format.

Sacraments

During the coronavirus pandemic, Grace United Church of Christ offered "virtual communion" on first Sundays. Each person provided his or her own elements, which typically consisted of a cup of water or juice and a cracker or a piece of bread. Participants were notified at the beginning of the service to have these items ready so that all could partake together.

The service followed a set sequence:
- Pastor introduced the Lord's Supper.
- Pastor led the congregation in "Confession of Sins:" "Most merciful God, we confess that we are in bondage to sin and cannot free ourselves. We have sinned against you in thought, word, and deed, by what we have done and by what we have left undone. We have not loved you with our whole heart. We have not loved our neighbors as ourselves. For the sake of Jesus Christ, have mercy on us. Forgive us, renew us, and lead us, so that we may delight in your will and follow in your ways to the glory of your name. Amen."
- Pastor offered assurance of pardon.
- Pastor broke the bread and dedicated the cup, as she blessed the elements for everyone.
- Pastor spoke the words of institution as the worshipers ate and drank the elements together online.
- Pastor offered a Prayer of Thanksgiving.

Throughout, Grace UCC was always mindful of the sacredness of the Eucharist ritual. Holy Communion involves the act of eating and drinking together to commemorate the death and resurrection of Christ. As often as we eat the bread and drink the cup, we do so in remembrance of Jesus, our risen Lord.

Cyber Outreach

Giving

During the pandemic, the church's program of giving continued virtually. The Pastor emphasized faithfulness in support of work of the Kingdom. She noted that giving requires the highest standard of personal integrity and discipline. It is not a casual or occasional thing. In 2 Corinthians 8-9, the Bible teaches us to give willingly, generously and cheerfully.

The Pastor emphasized that our giving pattern should reflect a regular and systematic plan. In Malachi 3:10, we learn God's standard for the tithe: the first tenth should be set apart from all earnings as a gift to God. Through our tithes and offerings, we prove faithful in the stewardship of all God has provided for us.

With these two principles in mind, the opportunity to give online was available to expedite one's personal plan for giving. Congregants and others were able to contribute through a number of channels, including PayPal, BillPay, Cash App:$GUCC2500 or they could mail a check to the church.

Membership

Even during the pandemic, the doors of the church were opened for membership. At the end of each sermon, the Pastor extended an invitation to listeners who were without a church home and interested in becoming a member of the Grace United Church of Christ family. Through the church's website, interested persons could simply complete a form online, and someone from the church would contact them.

Technical Considerations

It goes without saying that Grace UCC, like other congregations, had to make adjustments in order to keep the church vibrant. As noted earlier, two of the things we did included using Zoom for Bible class, and livestreaming Sunday morning worship service. Many of the challenges the media coordinator faced in successfully executing these initiatives pertained to technology. The following paragraphs describe what Grace learned from their efforts.

Zoom Bible Study Broadcasts

Connectivity, sound quality, and recording of the sessions were some of the problems Grace encountered frequently. Below, the media coordinator provides useful advice on how to avoid problems.

1. Whenever possible, meeting facilitators must connect to Zoom calls using computer audio and ethernet.
2. Whenever possible encourage all participants to use headphones and an external microphone. This helps to eliminate feedback.
3. Whenever possible participants should turn off any fans within the room to avoid the low level noise that will cause audio feedback in the Zoom meeting.
4. If you're going to record a Zoom meeting for live streaming, whenever possible, the meeting host should record the zoom meeting on their local workstation as opposed to recording in the cloud. There are two reasons for this:

Cyber Outreach

   a. Whenever you are recording to the cloud, there is a risk of stream interruptions that are reflected in the recording.
   b. When Zoom sends you the recording, they send you the file at the lowest resolution. In Grace's case, that file needed to be enlarged to match the resulting broadcast which created some artifacts in the recording.

Livestreaming Worship Services

For the five years prior to the mandated sheltering of churches, Grace UCC had used digital technology to routinely record sermons and to broadcast the recordings later on the church's social media platforms. The media coordinator started livestreaming to Facebook, and over the years expanded to YouTube, Twitter/Periscope, Instagram, and the church's website. These are some of the issues that the media coordinator encountered.

1. The quality of the livestream is heavily dependent on the computer hardware, software and cameras used to record the sermons.
2. Video is very resource dependent. By this we mean video requires a lot of memory and cpu speed to edit and encode. We started using a MacBook Air mid- 2011 with 4 gigabytes of memory. We quickly, changed to a Mac Mini with 8 gigabytes of memory. At some point, we began uploading the edited video to Amazon Web Services for encoding.
3. When we started recording, we used USB webcams mounted on tripods. We now use a Sony DSLR 35 mm camera that can record in high definition resolution. We currently broadcast the sermons at a resolution 1280 pixels

wide by 720 pixels high. The webcams simply could not match the quality of the DSLR cameras.
4. Lights are very important. We installed sufficient lights to minimize shadows, exposure, and hue issues in the recordings.
5. Sound is also very important. We have experimented with many configurations and currently use a combination of front-of-house audio feeds and direct sound captured by condenser microphones.
6. It is extremely important to have a reliable Internet Service Provider that provides high speed ethernet connections with sufficient bandwidth to handle video streaming.
7. We also began to use a restreaming service which allows the broadcast to be streamed concurrently on multiple platforms. We use the same restreaming service for livestreaming both the Zoom meetings and the Sunday Worship Service. (We also discovered that the restreaming service has their own video encoding requirements that your video must meet.) Simply put, we now upload or stream directly to the restreaming service. The restreaming service then takes that livestream and transcodes in real time to each of the social media platforms that we select. Each social media platform has their own streaming requirements. Our restreaming service is a Fortune 500 corporation with servers worldwide. We learned to trace the network paths from our church to the restreaming service servers. We also have encountered situations when the intermediate servers were crashing, overloaded or under attack by hackers, which resulted in delaying our stream. We describe these challenges to emphasize that the selection of the restreaming service is a critical decision.
8. Computer Software. We elected to use Final Cut Pro as the primary software to edit our videos. Final Cut Pro is an Apple product that uses Motion and Compressor as companion pieces for video editing and encoding. Final Cut

Cyber Outreach

Pro has a large third-party selection of plugins, titles and transitions that can be purchased and used. One of the lessons we learned is to review those plugins to ensure that they work in your unique computer hardware configuration. As the plugins get more sophisticated they require more resources to run properly. Also, as the Mac Operating System evolves, some of the plugins will not run on the updated software.

In the next chapter we describe how to lay the groundwork for the success of your digital ministry. We chose to use strategic planning as a framework because it represents a time-tested process for establishing programs in a range of settings. The process translates well to church settings, just as it would for any complex organization. We discuss the importance of establishing supportive environments for success. We also describe the type of leadership that would be most effective in directing a structured program that accounts for the needs of the people involved in the organization.

Chapter 5: Planning for Success

A Strategic Planning Process

Strategic planning is a process that entails an examination of a ministry's values, current status, and environment, and then relates these factors to the organization's desired future state. The process addresses three questions: (a) Who are we? (b) Where are we going? (c) How will we get there? The planning process begins with a belief statement and a vision of what the community should be, then provides a framework that guides choices related to the future nature and direction of the community. Through strategic planning cyber ministries can define a basic mission then allocate resources to be devoted to accomplishment of the mission. The end product is usually expressed in terms of a five-year or ten-year plan.

Another way to look at strategic planning is to view it as a systematic process to assess a church community's steadfastness in carrying out the stated mission and advancing toward a common shared vision. Using input from the internal and external community as a barometer, the process provides new, and perhaps better, information to support decisions related to sustained growth. A number of differences exist between conventional planning and strategic planning. Patterson et al. have developed a table to contrast the two.[27]

Table 6. Differences in Planning Types

Conventional	Planning Strategic Planning
Internal World View	External World View
Segmented Perspective	Integrated Perspective
Long-Range Planning Horizon	Medium to Short-Range Planning Horizon
Quantitative Data	Qualitative Data
Master Plan	Masterful Plan

The purpose of strategic planning is to help an institution capitalize on its strengths, its weaknesses, to take advantage of opportunities, and to defend against threats. In strategic planning, the role of the community is examined within the context of its environment.

Mission/Vision Driven Focus

Every viable plan for change should include a clearly defined mission and vision that serve as the driving force behind everything that takes place in your ministry. They serve as the umbrella under which all other activities will occur. Your mission and vision constitute the mutually defined, commonly shared purposes for the existence of the ministry, with a future oriented sense of what the community is capable of becoming. Together, the mission and vision provide coherence for the work you will accomplish.

Mission

The ministerial leader must work with others to define the mission statement, which identifies the scope of the faith community's operations, reflecting its values along with its outreach priorities. One major purpose of a mission statement is to help them make consistent decisions, to motivate, to build unity, to integrate short-term objectives with longer-term goals, and to enhance communication. The mission statement must have a target audience that ministry committees will keep in mind as they draft components of the mission. In a practical sense, the mission statement addresses three key questions:

- "To what do we aspire?"
- "What is our dream?"
- "What is our purpose?"

A ministry's mission statement succinctly sets forth the core values of the digital community. In a global way, it communicates to everyone that which is important and indicates the direction the ministry will take. It is the objective towards which the whole community is moving. It is the focus of everyone's actions.

The mission statement must also answer another important question that reflects the program focus of the ministry: "What do we want our parishioners to know and be able to do as they take part in this cyber ministry?" The most straightforward answer to this question ought to directly address the programs and services the ministry provides to parishioners.

A good mission statement is relatively brief and easy to remember. It provides direction for all parties involved regarding decision-making and resulting actions. It also informs the community of the commitment of the cyber ministry to improving the spiritual well-being those involved.

Here are a few other characteristics that a mission statement should include. For example, the format of the mission statement may be left up to the creativity of the cyber ministry planning committee. The document might take a form that is suitable for framing, for reproduction on wallet-sized cards, or for anything else in between. The mission statement itself should be just long enough to reach the target audience, that is, anywhere from one sentence to no more one page in length. And although any statement of the community's mission should be designed to last many years, the committee should not be afraid to rewrite its mission when it becomes necessary to do so because of changing circumstances. Language used in the mission statement should be chosen deliberately and should appeal directly to the target audience.

Vision

In the process of establishing a mission, the cyber ministry planning committee should also develop a community-wide vision that addresses the long-term hopes and dreams of what the ministry can become. The ministerial leader might start by asking community planners to brainstorm about the characteristics of the ideal cyber ministry. Ask them to compare their present status to the ideal. In doing so, ministry planning committee members are able to identify a number of gaps between the real and the ideal. Further discussions will lead to identification of ways to bridge the gaps.

Establishing Supportive Environments

In order to facilitate growth toward goal accomplishment, attention has to be given to the culture, capacity-building, and structural components of the cyber ministry. Such an environment provides abundant opportunities for members of the cyber community to become agents, rather than objects, in matters pertaining to their spiritual lives and their work.[28]

Culture

An appropriate culture provides opportunities to build relationships, establish mechanisms for reinforcement and affirmation, and maintain frameworks for social, spiritual, and psychological safety.[29] The culture should be nurturing, safe, confidential, supportive, and validating to help cyber community members formulate their ideals and shape their identities in ways that are empowering and liberating.

In many respects, churches, whether online or on ground, are "storehouses of our memories."[30] Affecting any type of change in what we know traditionally as "church" or other sacred communities such as schools, must involve a fundamental renegotiation of "cherished myths and sacred rituals by multiple constituencies."[31] When speaking about sacred communities, Terrence Deal further said, "The entire community must reweave or reshape the symbolic tapestry that gives meaning to the [growth] process, and this takes time"[32]

Moreover, within each cultural framework exists a set of phenomenological perspectives, which Thomas Sergiovanni refers to as "mindscapes." These implicit intellectual and psychological images establish the framework through which parishioners envision reality. Sergiovanni says that mindscapes set the boundaries and parameters that help people make sense of the world.[33] They provide rules, assumptions, images, and serve to shape thinking and belief structures about ministerial study and practice.

Capacity

Capacity equates to opportunities to practice and hone new skills. Building capacity within your cyber ministry will entail providing opportunities for community members to participate meaningfully in the programs offered. Capacity building may also involve helping community members engage in activities that will help them become self-assured in their faith commitments and ardent voices for justice in the world. Such activities might include:
- Shared norms and values
- Collaboration
- Reflective dialogue
- Collective focus on problem solving
- Spirit of shared responsibility for each other
- Applied scriptural analysis
- Personal and spiritual empowerment
- Critical consciousness about the issues of the day

Structure

Supportive structures are evident in the way each ministry is organized. Acceptable structures would be free from technical

glitches and complicated access procedures. In addition, the technology used should facilitate a supportive online community that includes:
- Openness to improvement
- Trust and respect
- Accurate knowledge
- Supportive leadership
- Technology support
- Clear communication plans
- Mutually convenient meeting times
- Adequate financing
- Other

Ministerial leadership

Leadership is another critical function impacting the success of any ministry. According to Newport, "There is little doubt that outstanding church leadership can be a powerful factor in facilitating the degree to which members feel closer to God, learn how to become better people, and get comfort in times of trouble and sorrow."[34] Outstanding ministers will be those who have a warm and caring attitude toward their parishioners and who can manage the affairs of the church or ministry for which they are responsible. Servant leadership provides a model for such management.

Servant Leadership

Servant Leader theory posits that the most effective leaders are those who are willing to be servants first, leading from a desire to better serve others and not to attain more power. Robert Greenleaf has asserted, "Servant leadership is a philosophy and set of practices that enriches the lives of individuals, builds better organizations and ultimately creates a more just and caring world.[35] To that end, Greenleaf has identified 12

Characteristics of Servant Leaders,[36] which are summarized below for your convenience.

12 Characteristics of Servant Leadership

1. Listening: Any successful traditional leader will tell you that communication and decision making are important factors in their ability to influence their constituents in a positive way. Servant-leaders are no different. They, too, must convey a commitment to listening to others intently. Active listening helps the leader "identify and clarify the will of a group." Therefore, servant-leaders listen receptively to what is said as well as to what is done by others. Greenleaf has reported that "listening also encompasses getting in touch with one's inner voice, and seeking to understand what is being communicated."[37]

2. Empathy: Understanding and empathizing with others is another behavior that servant-leaders exhibit. They try to look for the special and unique characteristics of others. They make so-called spirit-to-spirit connections with those whom they serve. Having empathy also entails assuming "the good intentions of employees/partners and not reject[ing] them as people, even when forced to reject or call into question their behavior or performance."[38]

3. Healing: Greenleaf noted: "One of the great strengths of servant-leadership is the potential for healing one's self and others." Healing is essential if growth and renewal, integration, and transformation are to take place. In *The Servant as Leader*, Greenleaf wrote, "There is something subtle communicated to those being served and led if implicit in the compact between the servant-leader and led is the understanding that the search for wholeness is something that they have."[39]

4. Awareness: This characteristic pertains to self-awareness as well as awareness of the needs of others. Servant-leaders who are aware have a strong sensitivity for what is going on. They scan the environment continually to note the subtleties and

nuances that occur. "They are always looking for cues from their opinions and decisions. They know what's going on and will rarely be fooled," according to Greenleaf.[40] In doing so, servant leaders are able to respond more readily when action is required.

5. Persuasion: Servant-leaders cultivate skill in persuading others, rather than relying on positional authority or coercion in making decisions. Servant-leaders convince others to comply with what is needed. Having the ability to persuade is "one of the clearest distinctions between the traditional authoritarian model and that of servant-leadership," according to Greenleaf. Having the skill to persuade becomes important when it is important to build consensus within groups.[41]

6. Conceptualization: Servant-leaders are people who "dream great dreams," according to Greenleaf.[42] This involves looking at the organization and the issues that emerge and conceptualize fresh approaches to problem-solving. Greenleaf calls this thinking beyond the day-to-day realities. At the same time, servant-leaders have to be able to strike a delicate balance between conceptualization and day-to-day focus.

7. Foresight: Servant-leaders are visionary leaders who have the ability to "understand lessons from the past, the realities of the present, and the likely consequence of a decision in the future. It is deeply rooted in the intuitive mind," according to Greenleaf.[43]

8. Stewardship: Servant-leaders with a strong sense of stewardship are those willing to prepare the organization to contribute to the greater good of society. Thus, they are responsible for "preparing it for its destiny," according to Greenleaf.[44]

9. Growth: Servant-leaders take responsibility for helping people grow. They believe that all members of the organization have something to offer beyond their tangible contributions. Servant-leaders, then, make concerted attempts to connect to

the developmental needs of others and actively find ways to help them reach their true potential.[45]

10. Building Community: Servant-leaders possess and convey a strong sense of "community spirit and work hard to foster it in an organization." Believing the organization should function as a community, they work hard to build cohesiveness from within. Greenleaf asserts, "Servant-leaders are aware that the shift from local communities to large institutions as the primary shaper of humanity has changed our perceptions and caused a sense of loss." Accordingly, servant-leaders try to find the means for building community among those who are part of the larger organization.[46]

11. Calling: Servant-leaders have a natural desire to attend to others and are willing to sacrifice self-interest for the good of the organization. According to Greenleaf, "This notion of having a calling to serve is deeply rooted and values-based."[47] Leaders who are also servants desire to make a difference among others in the organization and do their best to pursue opportunities that impact the lives of everyone in the community. They never act solely for their own gain.

12. Nurturing the Spirit: Servant-leaders are people who nurture the spirits of those in the organization. They accomplish this by praising organizational members honestly and supporting them by recognizing their efforts. When the servant-leader must give criticism, the criticism is offered without harshness. Greenleaf observed that the servant-leader "reminds employees to reflect on the importance of both the struggles and successes in the organization and learn from both."[48]

In the next chapter we provide a step-by-step guide for strategic planning. Such planning begins with assessments of both the external and internal capacity of the organization. We emphasize the importance of a belief statement and give guidance on how to analyze the gap between the real and ideal

Cyber Outreach

goals of the outreach undertaking. We describe action planning and provide detailed information to help you develop an action plan. We also discuss the importance of the budget and provide a template for a budget proposal.

Chapter 6: Steps in Strategic Planning

The literature varies on the precise number and nature of components that should be included in strategic planning. The following list encompasses the most commonly cited steps in the process:
1. Assessing the external environment
2. Assessing internal capacity
3. Defining the belief statement, the vision, and the mission
4. Establishing goals, objectives, and action plans
5. Identifying funding sources
6. Implementing the plan
7. Assessing and evaluating the plan

Assessing the External Environment

Assessing the external environment can be accomplished in many different ways. During this phase of the process, planners typically identify threats to accomplishing the mission and pinpoint opportunities for collaboration among members of the community. Among the factors that should be examined are the economic, demographic, social, technological, and political concerns that may have implications for the community's future. Methods for gathering information about the external environment include surveys, interviews, the key informant technique, community forums, and other mechanisms.

Assessing Internal Capacity

Each community, being unique, has to engage in planned self-examination to sort through the varying perspectives that parishioners and supporters bring to the table. All too often a factor identified as a strength by one community member may be identified as a weakness by another. During the internal analysis, planners will identify strengths and weaknesses existing in the targeted cyber community of faith. Factors to be examined include church governance, culture, leadership, staff expertise, commitment, Christian educational curriculum, and congregational characteristics.[10] It is important for church cyber ministries to reach consensus on key factors in these areas, since values underlying the associated practices reside at the heart of the community's culture and, therefore, have an impact on success.

A needs assessment will help ministry committee planners to define the scope and nature of the outreach. During the assessment, the community planners should observe, interview, and/or survey parishioners and community members. In addition, the planning team will conduct informational meetings to outline goals, objectives, and activities for incorporation into an integrated action plan.

The action plan will detail observable and measurable activities in which the cyber ministry will engage to address the objectives that have been identified. The plan will also specify the time frame in which the activities are to occur, the persons responsible for coordinating the activities, the indicators of successful accomplishment of objectives, and the costs associated with each activity.

The church governing board should have an opportunity to review the draft action plan and to provide their feedback before final approval. Keep in mind that it usually is not enough to assume that everyone shares a common understanding of where the ministry is headed and how it is going to get there.

Evaluation should be ongoing. Some elements of the evaluation plan will include conducting monthly reviews to assess "what's working" and "what needs to be changed," establishing opportunities for ministry coordinators to attend professional development workshops or webinars periodically to share success stories and learn about best practices, and to celebrate successes as a group periodically.

Other evaluation measures will include informing the community on an ongoing basis (at least twice a year) of positive growth, publishing a periodic newsletters that would be sent to parishioners, obtaining feedback from the community at large through surveys, maintaining a "suggestion box" to receive input from the community, and inviting members of the community to participate in quarterly forums to discuss progress.

The Gap Analysis

Together, these components drive the strategic planning process and serve as the central point of focus for all developmental activities. Following the evaluation, the community should conduct a gap analysis. The gap analysis forces planners to examine the space separating the ideal from the real and to identify ways of bridging the gap between the two. The gap analysis is a fairly simple process. We provide a sample gap analysis template in Table 7.

Table 7. Gap Analysis Template

Area Of Analysis:				
Goal For Analysis:				
Ideal (desired situation)	Real (Existing situation)	Gap in real and ideal	What is needed to reduce gap?	Issues and risks

Data gathering during this phase will entail the use of interviews, observations, questionnaires, content analysis, and quantitative data analysis techniques. The gap analysis document can also be used to track simple goals. Pinpointing patterns in the data will reveal critical issues, which will then serve as the basis for establishing goals, objectives, and action plans. A sample action plan is included in Table 8.

Table 8. Action Planning Template

Tasks/Steps Required to Accomplish Goal	People Responsible for Task Completion	Materials Needed	Funds and Sources	Date Started	Date Finished
Goal 1:					
Goal 2:					
Goal 3:					

Timeline

Task	Begin	End	Responsible

Budget

Item	Explanation	Amount
Travel/ Lodging		
Contractual Services		
Equipment		
Software		
Data Entry / and Clerical		
Telecommunications		
Training and Education		
Other Commodities		
Indirect Costs		

Establishing Goals, Objectives, and Action Plans

Setting goals and objectives is the next step in the process. These will be linked closely to the beliefs, vision, and mission of the cyber community of faith. Goals will emerge by examining what is to be achieved and in what order. Specific objectives will grow out of this activity and will represent smaller, observable and measurable steps toward accomplishing the goals. Action plans can be even more detailed, in that they specify what will occur, who will be responsible, how will it take place, and when will it be done.

Identifying Funding Sources

The strategic plan should contain a budget that reflects priorities to help the ministry achieve the stated mission and realize the goals. The budget represents a commitment by the community to fulfill the provisions of the plan. It responds to the query, "How much of our time, talent, testimonies, and treasures will we need to accomplish the mission?" The budget translates the community's priorities into financial and action terms within the context of available resources. Once it has been approved, the budget establishes the basis for all spending within the cyber ministry community. Table 9 shows a sample of a proposed budget.

Implementing the Plan

Most of the activities in the action plan are short-term in duration and can be monitored to ensure they are completed according to the time lines established. Not only that, the action plan identifies people who will take the lead on completing each action step. In the long run, if all steps are followed as planned or if adjustments are made to account for unforeseen events, the community will be able to move systematically toward its goals.

Table 9. Proposed Budget

Item	Explanation	Amount
Educational Materials	Curricular Supplies and Materials	$1,000
Support Personnel	Part-time Technical/Media Interns (2)	$6,000
Equipment / Technology	Cameras, Lights Space Enhancements Sound Servers and software Computers Website enhancements Runners for curtains Curtains	$10,000
Indirect Costs	Church Facilities Overhead Clerical Services Advertising Printing	$2,000
Total		**$19,000**

The remainder of this chapter provides a series of templates that cyber ministry planners may use to monitor implementation of the strategic plan. These same templates may also be used during the assessment and evaluation phases to determine how successful your program has been.

The first template is shown in Figure 5. It can be used for either grant or program evaluation. You will notice there is a place for you to record your vision, mission, goals and objectives along with a narrative summary of your project.

Narrative Summary

The narrative summary comprises a general overview of what you will accomplish and the tools you will use to determine whether or not you have met your objectives. The tools should comprise a mixed-methods approach. In other words, you will demonstrate accomplishment of your outcomes by providing both quantitative and qualitative indicators.

Sample Template for Grant Evaluation

Vision

Mission

Goals

Objective 1
   Observable and measurable statement of an objective. Be specific.

Objective 2
   Observable and measurable statement of an objective. Be specific.

Objective 3
   Observable and measurable statement of an objective. Be specific.

Etc.

Narrative Summary
General overview of what you will accomplish and the tools you will use to determine whether or not you have met your objectives. The tools should comprise a mixed-methods approach. In other words, you will demonstrate accomplishment of your outcomes by providing both quantitative and qualitative indicators.

What progress will you make toward reaching outcomes each year of the project? Table 1 displays the objectives and activities over the life of the project. Table 2 shows accomplishments for reporting year.

Figure 5. Sample Template for Program/Grant Evaluation

Cyber Outreach

The next graphic shown in Figure 6 displays a sample table you may use to record your progress toward meeting your objectives and completing associated activities. For each year of the project, you will list the objective and activities you performed to accomplish each objective. To document your accomplishments, you should include evidence of completion in the form of surveys, interviews, stories, videos, etc. Although Figure 6 provides space for three years of activities, you may use this template to keep track of your progress for a single year, if that is your preference.

Progress toward Meeting Objectives
Table 1. Timetable – Activities over the Life of the Project

Tasks	Evaluation Year 1	Evaluation Year 2	Evaluation Year 3
Objective 1			
Activity	Surveys Interviews Stories Videos	Surveys Interviews Stories Videos	Surveys Interviews Stories Videos
Activity			
Activity			
Objective 2			
Activity			
Activity			
Activity			

Table 2 provides an overview of accomplishments for each year. Details for the activities are described in the narrative that follows the table.

Table 2. Accomplishments for Year 3

Proposed Task	Activities/Evaluation of Completion
Objective 1	1. 2. 3.
Objective 2	1. 2. 3.
Objective 3	1. 2. 3.

Figure 6. Progress toward Meeting Objectives.

Figure 7 can be used to record an overview of your program accomplishments for each activity. In the detailed narrative, you will describe specifically the nature of activities you completed to achieve each objective. It is always helpful to include indicators/evidence of success (i.e., survey results, interview summaries and quotes, anecdotes, videos, tables, figures, and/or photographs) to document your progress. In the section labeled summary and recommendations, you will include a brief, narrative recap of what you accomplished along with recommendations on how you might improve in subsequent years.

Detailed Narrative Description for Each Activity
Here you will describe specifically the activities that were completed for each objective. You must include indicators/evidence of success (i.e., survey results, interview summaries and quotes, anecdotes, videos, tables, figures, and/or photographs).

Objective 1.

Objective 2.

Objective 3.

Plans for Future Activities

Objective 1

Objective 2

Objective 3

Summary and Recommendations
This section will include a brief, narrative recap of what you accomplished along with recommendations on how you might improve for subsequent years.

Recommendation 1:

Recommendation 2:

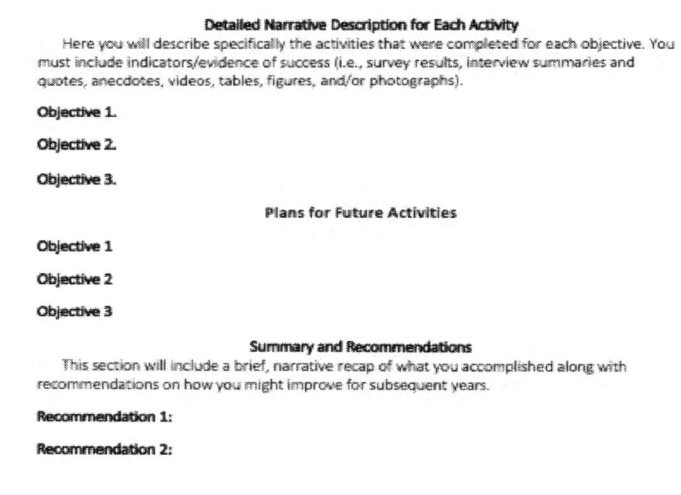

Figure 7. Detailed Narrative for Each Activity

The next chapter will focus on how to assess the effectiveness of your digital ministry. We provide advice on establishing data-gathering procedures and administering data-gathering tools. We give examples of questionnaires and outcomes from prior evaluations we have conducted.

Chapter 7: Assessing and Evaluating the Plan

As an example of how to assess and evaluate your strategic plan, we are providing examples from a case study of the Christian Education Department at Grace United Church of Christ. We used survey research, which, by definition, entails the collection of descriptive, quantifiable data pertaining to questions concerning the current status of participants in the study. Creswell reported that the survey design allows the research to construct a quantitative description of trends, attitudes, or opinions of a population by studying a sample of that population.[49]

Participants

The Christian Education Department study used purposeful sampling to identify participants to complete questionnaires. According to Patton, purposeful sampling allows the researcher to identify participants that will provide the basis for information-rich cases.[50] Also, with this kind of sampling, "the focus is on generating in-depth information and understanding of individual experiences" from a small number of people who have knowledge of the studied phenomenon.[51]

Data Gathering Procedures

In an effort to provide appropriate programming for parishioners, the Christian Education Department conducted a survey to assess the needs and interests of participants. The case study was conducted in two phases. Phase 1 involved review of archival data including annual and quarterly reports, meeting minutes, video tapes, and other existing artifacts along with data collected from a needs assessment survey in 2014. Phase 2 consisted of a follow-up structured questionnaire with some open-ended items administered in 2017. Program evaluation questionnaires were sent to all the people on the church's Mailchimp mailing list. The anticipated sample response size was n=30.

Data Gathering Tools

Questionnaires

The questions of the study were structured for the purpose of acquiring specific information about how well the Christian Education program was meeting the goals and objectives specified in the strategic plan. The researcher summarized the results from data collection using descriptive statistics and percentages to summarize questionnaire responses. Some of the data are depicted in the form of tables, charts, and graphs.

Other Data-gathering Tools. We further analyzed a variety of artifacts.
- Archival data
- Video tapes
- Audio Tapes
- Observation Notes
- Questionnaires (program evaluation, demographic, open-ended)
- Interviews
- Focus Groups
- Photographs

Description of the Christian Education Program

Mission

The Mission of the Christian Education Department is to provide sound educational experiences based on the Word of God to empower God's people for transformational action in their lives, their communities, and the world.

Vision

The Vision of the Christian Education Department is to promote biblical literacy and application of biblical principles of discipleship. These principles will be presented and reinforced through structured and non-structured educational programs, including Sunday school, Bible Basics classes, Thursday Evening Bible study, in addition to other thematic and applied courses of study offered on a regular basis. Classes will be taught online and face-to-face.

Goals

The overall goal of the Christian Education Department is to deliberately cultivate and effectively teach a Bible-based, culturally relevant curriculum designed to promote justice and transformation within the body of Christ and the community at large.

The primary audience consists of adult learners. In that regard, the Department seeks to nurture knowledge seekers through a variety of platforms, including face-to-face classes and virtual outreach. Table 10 provides Facebook metrics for 2015-2018. Through these channels, the Department will help seekers refine their biblical scholarship skills and engage in discernment about the meaning and application of an ancient text in a modern world.

Table 10. Thursday Night Bible Study Facebook Metrics (2015-2018)

Year	Reach	Impressions
2015	14,224	30,121
2016	160,120	275,878
2017	130,871	242,161
2018	173,941	195,546

After reviewing results of the surveys, we made changes in programming for Christian Education. Results of the studies were not generalized beyond the church population. Out of this analysis, our primary program goals emerged. We used survey data as a guide to formulate additional recommendations for change.

Goal #1: Target Higher-Level Adult Learning

As the result of involvement in structured online learning experiences, participants are expected to acquire and refine a practical skill set for living righteously in the present day. This includes problem solving, analysis, spiritual empowerment, and critical consciousness about the linkages among their faith, discipleship and social justice in contemporary settings.

Objective 1: By the end of 2020, participants will read scripture to discern its meaning for the original audiences.

Objective 2: By the end of 2020, participants will discern how scripture can be applied to contemporary situations.

Objective 3: By the end of 2020, participants will lead a Bible study class from beginning to end.

Goal #2: Continuation of Adult Sunday School

In its configuration prior to the surveys, the on-ground adult Sunday school provides a rich forum for discussion and application of Scripture. Approximately 12-15 people attend the classes, which meet in person from 9:30 a.m. to 10:15 a.m. each Sunday morning. For the past several years, the Christian Education Department has used curriculum materials developed by Urban Ministries International (UMI). The materials provided a framework for helping learners see themselves in the biblical text and apply its essential principles to the contextual circumstances of their lives. The Sunday morning educational encounter encourage adult learners to draw on and share the challenges, experiences, and insights gained during their journey as Christians. This approach allows collective wisdom to grow out of the community's experiences as key issues are discussed and unpacked. Survey respondents were satisfied with this Sunday school configuration. It was temporarily suspended, however, due to the sheltering in place orders imposed as a result of the pandemic.

Objective 1: Participants will plan and lead a Sunday school lesson using UMI curriculum materials.

Objective 2: Participants will share stories to describe the challenges, experiences, and insights gained during their journey as Christians

Objective 3: Participants will lead prayer at the beginning and/or end of the Sunday school lesson.

Goal #3: Cyber Adult Christian Education

As a result of the surveys, the Department decided to expand its livestreamed broadcasting of Thursday Night Bible Study classes. Videos of the classes are available for viewing by the public on demand through the church's social media platforms, inhouse archives, and YouTube. Broadcasting via the Internet facilitates ongoing outreach to a larger audience. On an average, approximately 90 people view these programs on a regular basis.

Objective 1: By the end of 2020, participants will lead Bible Study classes using Zoom or similar technology.

Objective 2: By the end of 2020 participants will develop lesson plans, which will include the following components:
- Discuss how biblical principles were applied for original readers of the text
- Engage in critical reflection about biblical topics
- Evaluate contemporary situations in light of biblical principles
- Exhibit confidence in their ability to share their faith with others

Objective 3: In order to facilitate outreach to a larger virtual audience, by the end of 2020, the Thursday Night Bible Study class will be available for viewing by the public via the

Internet on demand through the church's social media platforms, inhouse archives, and YouTube.

Outcomes

As the result of the evaluation, we revised our Thursday Night Bible Study schedule to accommodate new interests (see Table 11). Different facilitators volunteered to lead the sessions noted below. All sessions were videotape via livestream and made available later for viewing by others in the Internet. We believe these goals conform to the Great Commission (Matt. 28:19-20). In addition, they equip those who hear the Word with skills for living in challenging times.

Table 11. Tentative Bible Study Schedule 2017-2018 (Revised 9/7/17)

Date	Topic	Length
9/7-10/12	*Galatians* Materials: Bible and Handouts	6 weeks
10/19	Movie Night - TBD	
10/26-11/16	1 and 2 Thessalonians Materials: Bible and Handouts	5 weeks
11/23	Thanksgiving Break (No Class)	
11/30	Celebration	
12/7-12/21	Advent Bible Study Materials: *Mary Had a Baby: An Advent Bible Study Based on African American Spirituals* by Cheryl A. Kirk-Duggan and Marilyn E. Thornton	4 weeks
12/28	Christmas Break (No Class)	
1/4-2/8	Black History (TBD)	6 weeks
2/14	Ash Wednesday	

Date	Topic	Length
2/15-3/15	Lenten Bible Study: Materials: *He Chose the Nails Study Guide: What God Did to Win Your Heart* (Paperback) by Max Lucado (Author)	5 weeks
3/22	Movie Night	
3/29	Maundy Thursday (TBD)	
4/1	Easter Sunday	
4/5-5/31	The Psalms Materials: *Answering God: The Psalms as Tools for Prayer (Paperback)* by Eugene H. Peterson (Author	9 weeks
6/7-6/14	Vacation Bible School (TBD)	2 weeks
6/7-7/26	Adult Bible Study: *Rest in the Storm: Self-Care Strategies for Clergy and Other Caregivers* (Paperback) by Kirk Byron Jones (Author)	8 weeks
8/2-8/30	No Bible study during the month of August	

Proposed Budget

At the time of the evaluation, the Christian Education Department broadcasted Thursday night Bible study classes via livestream using borrowed equipment. Because the equipment is not made for heavy-duty or professional use, the quality of the broadcasts is often uneven, and technical problems occur on a regular basis. The budget for the revised program reflects purchase of equipment for the digital ministry and training for others in how to operate the equipment and provide ongoing support. In order to meet the goal and accomplish the objectives of this revised project, the Department proposed a budget of $19,000 for the first year. Results of the study were not be generalized to other groups, but were intended to serve as a guide to formulate recommendations for change.

Chapter 8 proposes a model for a hypothetical program implementation using an action research approach. In this chapter we describe steps for using action research as a tool for ongoing assessment of the effectiveness of your ministerial outreach program.

Chapter 8: Possibilities and Potential

Active Voice Initiative Draft Proposal

We have included in this book a draft proposal for a cyber-based, collaborative project involving church and community. This chapter will give you an idea of how to conduct action research to address the needs and concerns of your parishioners and other members of the community at large. We believe that when church and community collaborate on solutions to common problems, even if these collaborations are digital, the outcomes have a greater and longer-lasting impact.

We call this project the Active Voice Initiative (AVI). We contend that the church is in a position to collaborate with community stakeholders to formulate solutions that account for not only the physical and emotional needs of the people, but also their spiritual wholeness. With church and community working together all parties involved can engage in problem solving on their own behalf. What these stakeholders need is a way to identify critical areas of concern, clarify the focus of their investigations, and implement solutions that are driven by the social, political, and economic realities endemic to their particular set of circumstances.

Collaborative action research models are appropriate for examining a range of complex issues in our churches and in society at large, issues such as violence, community renewal, church vitalization, systemic inequities, racial justice, and other areas. Because a range of divergent perspectives from within the community and church are brought to bear on the issues under investigation, the actions that emanate from collaboration have the potential to affect meaningful change within the community ecology.

In this model, stakeholders and others take charge of their own outcomes in ways that engender hope and self-determination among all community members. The practical skill-set that grows out of this approach includes problem solving, critical inquiry, soul searching and development of critical consciousness about how to change the circumstances of those who wish to live righteously.

Purpose of the Research

The primary purpose of this mixed-method, multidisciplinary, collaborative action research project is to equip church parishioners and community stakeholders with empowerment skills for making social, spiritual and economic change in their lives and in their communities. Specific project goals are to:

1. Teach collaborative action research skills to church parishioners and community members in an online environment.
2. Improve skill levels of participants in the following areas:
   a. Collaborative problem solving
   b. Data-driven decision making
   c. Information analysis
   d. Critical thinking
3. Make an impact on attitudes towards:
   a. Change
   b. Justice
   c. Resiliency
   d. Self-efficacy
   e. Spirituality
   f. Redemption
   g. Righteousness
4. Document the process for replication in additional settings.
5. Disseminate results of the project for others to emulate.

Procedures

To look at this framework from another perspective, collaborative action research engages a group of people in systematic trial and error. The focus is on practice – how do we work with what we have to make our circumstances better than what they are now? How do we glean from our history and our surroundings the information we need to make better decisions about how to live righteously? How do we raise our awareness of what is appropriate and activate a plan to make it happen?

The collaborative action research process comprises only three steps: Look, think, and act.

Look

The problem-solving process begins with looking at prior research and theory pertaining to the problem. For our purposes, it also involves studying scripture to examine issues using a biblical frame of reference. The approach has to be holistic as it takes into account the complex interactions among dynamic social, cultural and spiritual variables. As a matter of fact, the whole process involves building a comprehensive picture using evidence from a number of sources and disciplines then examining wisdom that emerges from both the past and the present. Having constructed the picture, stakeholders are able to scrutinize it using a variety of analytical and biblical lenses and filters. Viewing the evidence in light of a particular context is also crucial, as the results are not be generalized beyond the specific setting under study. "Looking" involves the following:

- Build a Picture
- Start with a general question or problem
- Define the problem
- Describe the situation
- Gather relevant information through
    o Interviews

- o Observations
- o Documents
- o Records
- o Surveys
- o Perceptions of others

Think

Thinking is the next step in the action research process. In this step, stakeholders are charged with interpreting and explaining what they just uncovered. Here they address the questions: Why, What, How, Who, Where, When. As they explore and analyze, stakeholders look for explanations that reveal the contextualized nature of the problem. What is happening here? How or why are things as they are? How do these things apply to this particular situation? Steps are as follows:

- Interpret and Explain
- Utilize interpretive questions: Why, What, How, Who, Where, When
- Explore and Analyze
- Look for explanations that reveal the contextualized nature of the problem
- Conjecture: What is happening here? (Hypothesis)
- How or why are things as they are? (Theorize/Theologize)

Act

In the third phase of this church/community collaborative is to act. Here, the stakeholders identify action steps that need to be taken in order to resolve the issues that have been identified. These steps will be recorded in an action plan. Staying on course involves giving some thought to the specific concerns that need to be addressed and how stakeholders intend to deploy resources to resolve problems under investigation.

At this stage in the process, stakeholders have to be intentional as they identify the tasks, steps, people, places, times, materials, and other resources to make things happen. Staying on track will require stakeholders to evaluate their progress every step of the way. Each day stakeholders should ask: Are we doing what we said we would do? If not, how do we get back on track? If necessary, stakeholders should discuss these questions with relevant outsiders or confer with experts if the issue lends itself to such conversations. To recapitulate, the action phase requires researchers to:

- Plan
- Review issues
- Organize issues in priority order
- Rate issues according to degree of difficulty
- Define tasks required to resolve the problem
- Formulate objectives (observable and measurable)
- Chart the course
- Identify task, steps, people, place, time, materials, funds
- Implement
- Carry out activities
- Engage in ongoing reflection
- Analyze activities
- Review information collected
- Evaluate
- Confer with outsiders or experts
- Assemble evidence
- Sort and interpret data
- Ask follow-up questions:
    o What new knowledge have we gained?
    o Who have we shared it with? (Validation)
    o What are our next steps?

In summary, the general characteristics of collaborative action research include:
- Collaboration among stakeholders

- Focus on real problems
- Date-driven decision making
- Change in practice resulting from new awareness
- Ongoing data collection and refinement

Conducting Action Research in the Online Environment

It is possible for the entire research process to be conducted online. Meetings could take place via Zoom or similar online platforms. Participants would receive preliminary materials through email prior to the meetings. It would be beneficial for participants to complete an online inventory to assess their preparedness for the upcoming activities. Also, if participants are uncomfortable with the technology, Zoom and other service providers offer free webinars for beginners to bring their skills up to par.

A range of presentation styles could be used to convey the information. These formats might include devotionals, lectures, small group chats, videos, and PowerPoint presentations. Variations in presentation formats would depend on the availability of equipment and other resources. Formats would also depend on the preferred learning styles of participants. Just to give you an idea of where people are with regard to their preferred ways of learning, we recently asked workshop attendees about their preferred styles. Their responses are summarized in the chart on the next page.

Conclusion: There is an Ongoing Need for Change

Hopefully, in this book, we have shown that times are changing, and the paradigm of the Christian church is in a state of evolution during the cyber age. There's been a shift in the way the Gospel message is delivered, particularly in the times of the recent pandemic. Given the current climate and readiness for change, it's time to make church (even in its digital configuration) relevant again. Ministerial leaders should ask, what are the costs involved in addressing change? A tandem question is, "What is the cost of doing nothing?" In an environment where everything has to be fabulously grotesque, extravagantly diverse, intensely scary, or extremely marginal, what should the 21st century church do to remain sustainable?

Preferred Learning Styles for Future Workshops

What is your preferred method for learning? (Mark all that apply.)

[Face-to-face classroom in physical proximity]	[Online learning]	[Independent supervised project]	[Self-guided video modules]
*			
	*		
*	*	*	*
*	*		
*	*		
*			
*			*

Until the answers to these questions become clear, trying to improve churches from the outside will remain little more than the practice of shooting arrows into the wind and hoping they land somewhere near the stated targets. We hope this book has demonstrated that what we need is strategic planning to accommodate meaningful change. We are responsible for developing a critical awareness of the impact of what we do on the outcomes we desire. And we must be ready to provide new answers to old questions. One size definitely does not fit all.

We submit to you that we need to continue to examine our priorities as the church moves towards alternative ways of outreach. It may be the case the 21st century priorities are slightly different from what they used to be. We offer several options to help you focus on new possibilities for establishing vibrant cyber ministries.

Creativity: The 21st century cyber church will use the Internet as a mission field to spread the Gospel of Jesus Christ through social networking and livestreaming religious broadcasts. As described throughout this text, the purpose of such programs is to edify the body of Christ and facilitate their movement toward spiritual maturity and dispensing of justice (Ephesians. 4:12-13).

Community: The 21st century cyber church will aim to teach a Christian curriculum based on the premise that learning is most effective when learners create personal meaning from new information; that is, they attempt to create a comfortable cognitive niche for the new information based on what they believe to be real and true in the world based on scripture.[52] Stated in another way, the learners will not be passive recipients of information but active participants in a process in which they develop meaning in a problem-solving mode, evoke meaning through interpretation of what is presented, and factor in the social and biblical context in which the information is relevant.[53] In that regard, knowledge is socially constructed and grows out of the understanding of each learner.

Justice: The 21st century cyber church will teach a curriculum using transformative learning,[54] a process based on theory that seeks to explain the dramatic changes in worldview, paradigms, and "meaning perspectives" among various people. Transformative learning theory uses a "disorienting dilemma" to prompt adult learners to reexamine previously held values, assumptions, beliefs, and perspectives. The discomfort of inner conflict compels them on a personal journey to make sense of contradictions in the world. This, in turn, prepares them to become agents of change and spiritual transformation. Using the language of Thomas Kuhn, transformative learning is similar to a "paradigm shift," or what occurs when a person undergoes a permanent change in the foundation of one's beliefs, values, commitments, and conduct.[55]

Spirituality/Faith: The 21st century cyber church will uplift spirituality and faith. During the Sunday cyber learning experience, learners will cultivate spiritual fitness. Guided activities will prompt learners to consider that our spirits are strengthened and liberated when we exercise regularly through worship, prayer, reflection, journaling, and other acts of devotion, whether it occurs online or otherwise. The curriculum will reinforce spiritual fitness, the idea that we are able to maintain faith in the face of mental and physical challenges. Spiritual wellbeing is evident in the person who finds meaning and purpose in life, and who operates from an intrinsic value system that guides their decisions. When learners are spiritually fit, they take on the mind of Christ.

Love: The 21st century cyber church will reinforce the value of love. Cyber worshipers will explore what it means to love God, neighbor and self in a world of relativity and individualism. While engaging the theme, they will press into service all the principles they have acquired as a result of their participation in a structured program of digital discipleship.

Summary

In summary, the 21st century cyber church will employ a range of ways to give depth and breadth to awareness and application of the basic principles of the gospel message, whether that message is promulgated in-house or online. We pray that this book has provided some food for thought as you move forward. We are providing here several of the steps you may follow to begin your outreach ministry:

- Establish your planning team
- Identify your mission vision goals and objectives
- Identify your target population
- Build community

- Select the media and platforms you will be using
- Establish a presence
- Send out meaningful posts
- Evaluate your results
- Make adjustments
- Refine your goals and objectives
- Repeat the cycle

This volume has provided basic tips on how to start your ministry and how to evaluate ongoing effectiveness. Based on the examples provided in this text, we hope you can see the possibilities in using the Internet for evangelism, discipleship, community building, prayer lines, external resource support, encouragement, cyber worship, preaching, teaching, research, and other forms of outreach. Ministerial leaders are limited only by their comfort level and creativity in using the World Wide Web. Moreover, ministerial leaders are led by the Holy Spirit. Given the current trend of declining attendance at brick-and-mortar churches and waning interest in institutional religion, the Internet represents a new mission field for reimagining the Great Commission.

In the appendices, we provide samples of some of the data gathering tools we have used or intend to use as we continue our investigation of effective ways to establish cyber ministry. You have permission to adapt these tools for use in your organizational setting.

Appendix A: 2014 Christian Education Survey Results

Question	Response
1. Do you **participate consistently** in bible study?	Yes = 6 No = 4
2. Do you think that bible study is **important?**	Yes = 10
3. What **day of the week** would be more convenient for you to come to bible study?	Mon. = 2 Wed. = 2 Thur. = 4 Sat. = 1
3. What **time of day** would be more convenient for you to come to bible study?	Morning = 1 Afternoon = 4 Evening = 3 No Pref. = 2
4. What would make you more **willing to participate** consistently in bible study, if you don't already?	Daytime hours Afternoons on weekends More convenient times Mornings except Fridays Whenever there is not another major obligation. Teleconference presentations Videos of sessions Multi-media Not sure

5. What kind of **topics** would you be interested in learning more about in bible study?

All books of Bible
A relationship with God/personal relationships
Acts of the Apostles
Walking by faith
Women of the Bible
African American presence in Bible
Bible stories
Bible journey from Genesis to Revelation
Study a character or book at a time
Bible basics
Application of Bible principles
No preference

6. Additional comments

I miss coming to Bible study, but it's really hard for me to get here. And I'm going to start trying to get to church early on Sunday for Sunday school.

I would love to participate in afternoon Bible study.

Appendix B: 2017 Christian Education Survey Results

The Christian Education Department gathered information to find out more about parishioners' interests in selected church activities. We used the results of this survey to improve our offerings and implement programs to meet the needs of community members.

1. Place a check mark beside the Bible topics you would be interested in studying. Mark only those that apply. N = 29

Topic	Percentage	#
Bible Basics-Introduction to the Bible	27.59%	8
Bible Study for Special Occasions (Lent, Advent, Epiphany, etc.)	31.03%	9
Church History	27.59%	8
Fasting and Spiritual Disciplines	48.28%	14
Overview of the Old Testament	41.38%	12
Overview of the New Testament	58.62%	17
Prayer Practices	58.62%	17
Self-Care and Physical Fitness	44.83%	13
Study of the Gospels	37.93%	11
Study of the Apostles	37.93%	11
Study of the Prophets	41.38%	12

2. Place a check mark beside the church-related activities you would like to be involved in. Mark only those that apply. N=27

Activity	Percentage	Number
All-Church, "One Read" Book Discussions	29.63%	8
Bowling League (or similar activity)	29.63%	8
Church-School-Community Partnerships	25.93%	7
Discipleship/Community Outreach	37.04%	10
Fundraising for Special Projects	37.04%	10
Job Skills Workshops	33.33%	9
Live-streamed Broadcasts of Bible Study	29.63%	8
Live-streamed Broadcasts of Sunday Morning Services	29.63%	8
Movie Nights (or similar gatherings)	44.44%	12
Stewardship/Financial Planning Workshops	25.93%	7
Workshops Dealing with Personal-life Issues	44.44%	12

3. Which of the following things keep you from taking part in Bible study? (Mark all that apply.) N=24

Item	Percentage	Number
I don't have the time.	29.17%	7
I don't have transportation.	0.00%	0
I don't know what classes are available.	12.50%	3
I'm not interested in any of the current topics.	4.17%	1
I don't feel I need Christian education.	0.00%	0
Other (please specify)	54.17%	13
TOTAL		24

4. Which of the following ministries/activities do you believe would be of interest to our children/youth? Mark all that apply. N=29

Ministries / Activities	Percentage	Number
Children's Church (Separate Youth Sunday Services)	41.38%	12
Christmas Pageant/Skit	48.28%	14
Creative Writing Ministry	24.14%	7
Easter Pageant/Skit	48.28%	14
Homework/Tutoring Ministry	48.28%	14
Leadership Training for Ministry	34.48%	10
Music Ministry/Youth Choir	62.07%	18
Research/Problem Solving Ministry	27.59%	8
Scholarship Fundraising Ministry	34.48%	10
Technology/Media Support Ministry	51.72%	15
Vacation Bible School	48.28%	14
Other (please specify)	10.34%	3

5. Which of the following methods appeal to you as effective ways to learn?

Learning Method	Percentage	Number
Lecture	24.14%	7
Book Study Discussion	55.17%	16
Group Discussion of Contemporary Issues	68.97%	20
Hands-on Experience	68.97%	20
Live-streamed Broadcasts of Bible Study	31.03%	9
Self-study	37.93%	11
Sermons	37.93%	11
Online courses	24.14%	7
Videos	31.03%	9
Teleconferences (i.e., Webinars)	24.14%	7
Other (please specify)	3.45%	1

6. When are you available to participate in Bible study or other church activities? Mark all that apply.

	Mon.	Tues.	Wed.	Thurs.	Fri.	Sat.	#
Mornings at Grace	28.57% 4	21.43% 3	7.14% 1	7.14% 1	14.29% 2	57.14% 8	14
Afternoons at Grace	33.33% 4	16.67% 2	25.00% 3	25.00% 3	8.33% 1	41.67% 5	12
Evenings at Grace	27.78% 5	38.89% 7	44.44% 8	50.00% 9	11.11% 2	11.11% 2	18
Suppertime/evenings at off-site locations	25.00% 2	37.50% 3	12.50% 1	37.50% 3	25.00% 2	37.50% 3	8

Comments: Showing **7** responses
- differs from week to week. Late mornings are usually better for me.
- Sunday
- I don't know right now.
- Mornings between 10am and 12pm Early afternoon between 1pm and 3pm
- Monday
- Na
- am--pm M-F

7.   Gender N=29

Female	68.97%	20
Male	31.03%	9

Cyber Outreach

8. Age Range N = 29

Age Ranges	Percentage	Number
Under 18 years of age	0.00%	0
18 to 29 years of age	0.00%	0
30 to 39 years of age	13.79%	4
40 to 49 years of age	17.24%	5
50 to 64 years of age	55.17%	16
65+ years of age	13.79%	4
TOTAL		29

9. Are you a member of The New Church?

Response	Percentage	Number
No	17.86%	5
Yes	82.14%	23
TOTAL		28
Comments Showing **3** responses		

- not official but I feel a part of the church.
- I am a member of [a nearby church] but come to Grace
- Calvary Community Church

10. Do your live in Sauk Village? N=29

Response	Percentage	Number
No	75.86%	22
Yes	24.14%	7
TOTAL		29

Comments Showing **6** responses *Chicago Heights, Chicago North Kenwood/Hyde Park, Park Forest, Lynwood, Hammond, Chicago

Appendix C: Sample Course Evaluation

Answer the following questions about the Bible study course you just completed. Rate each item using the following scale: 5=Strongly agree (SA), 4=Agree (A), 3=Somewhat agree (SM), 2=Disagree (D), 1=Strongly disagree (SD). Place a checkmark in the column that corresponds to the way you feel.

		SD	D	SM	A	SA
1.	I have a good understanding of what it means to be a disciple.					
2.	I can describe the context of a passage of scripture from the lessons.					
3.	I am able to describe at least one way scripture applies to contemporary Christian discipleship.					
4.	I am able to describe possible strategies for engaging in contemporary discipleship.					
5.	I can describe how biblical principles were applied for original readers of the text.					
6.	I am able to critically reflect on biblical topics presented in the lessons.					

7.	I have the ability to evaluate contemporary situations in light of biblical principles.					
8.	I am able to give examples of how Jesus served as a mentor to his disciples.					
9.	I can explain to the average person why discipleship is important.					
10.	I can give examples of how discipleship may be carried out in the 21st century.					
11.	I can describe what it means to "Listen for the Word of God."					
12.	I am able to reflect on marks of effective discipleship for the present day.					
13.	I feel comfortable discussing key issues from the scripture with other members of the class.					
14.	I am able to utilize outside resources (i.e., dictionaries, commentaries, concordances, etc.) for gaining new insights.					

15.	I can describe the differences among evangelism, discipleship, and spiritual formation.					
16.	I can describe discipleship looks like in today's world.					
17.	I am able to state my definition of discipleship in the 21st century.					
18.	I am able to discuss what prohibits discipleship in the modern day.					
19.	I can describe the background of the scripture as it relates to the original hearers of the text.					
20.	I can identify major themes/lessons learned from a passage of scripture.					
21.	I believe that using the Bible in weekly lesson facilitation helps class members apply the Bible to their local context.					

Cyber Outreach

22.	I am comfortable using outside sources to help me prepare to facilitate weekly lessons.				
23.	I enjoy being part of the Live Stream broadcasts on social media.				
24.	I believe that rotating facilitation of the lessons is a good way to help everyone become biblically literate.				
25.	I prefer using the Bible as the primary source for weekly lessons over using a textbook written by an outsider.				
26.	The course provided opportunities for participants to apply lessons learned to their individual contexts.				
27.	The length of the course was appropriate.				
28.	Overall, the presenters did a good job of facilitating the classes.				
29.	Overall, the course was very good.				
30.	I would like to continue using this format for classes in the future.				

How could the course be improved?
Other suggestions or comments
In what ways do you intend to use the information you learned in class?
Other comments

James and Lorrie Reed

Appendix D: Assessing Environmental Support Factors

To make sure we facilitate candidates' growth toward meeting program goals, attention has to be given to the structure, culture, and capacity-building components of the learning experience. The following may be used as a guide to facilitate discussion and exploration of these factors.

Structure

1. Does the overall **organizational structure** support accomplishment of goals and objectives?
2. Does the **program structure** support accomplishment of goals and objectives?
3. What structures enhance accomplishment?
4. What structures inhibit accomplishment?
5. What is the function of the governing board?
6. Does the governing board function effectively?
7. What are the benefits and detriments of online components of the program?
8. What supports are provided for the online components?
9. What are the strengths and weaknesses of the decision-making process?
10. What are the strengths and weaknesses of the communication process?
11. What are the strengths and weaknesses of the problem solving process?

Culture

1. Describe the culture of the ministry.
2. Provide evidence to support the existence of collaboration.
3. Provide evidence to support the existence of effective communication.
4. Provide evidence to support the existence of community.
5. Provide evidence to support the existence of growth in spiritual resiliency among participants.

Capacity

1. Describe how the curriculum promotes mastery of pertinent goals and objectives.
2. Evaluate the adequacy of the online academic program.
3. Evaluate the adequacy of the Thursday Night Bible Study.
4. Evaluate the adequacy of the Sunday Morning Prayer Calls.
5. Evaluate the adequacy of the Webinars.
6. Describe how leadership style enhances or compromises attainment of goals.
7. Describe how the curriculum promotes spiritual growth and development.

Appendix E: Sample Template for Social Media Metrics

	Reach	**Impressions**	**Engagement**
Facebook			
Instagram			
YouTube			
Periscope			
Other			

Appendix F: Internet and YouTube Outreach by Program

Activity	Reach	Impressions	Engagements
Preached Word (YouTube)			
Bible Study (TNBS YouTube)			
Music Ministry (YouTube)			
Facebook Page			
Twitter Account			
Periscope Account			
Instagram Account			

Appendix G: Program Evaluation Questionnaire

Purpose of the Study: The purpose of this questionnaire is to examine the progress of the Church in meeting its stated mission, vision, goals and objectives during its first five years of existence. The results of this study will provide evidence to support and reinforce the church's ideals. We would appreciate your assistance in this evaluation.

Section 1 Impact of Church Activities on Church Growth

For this part of the questionnaire, consider the following question - To what extent have the activities of the church had a positive impact on church growth? **Instructions:** Read the list of activities/accomplishments below. Reflect on them and assign each activity a grade of "A" through "F" based on your opinion of its effect on church growth. Rate the activity "NA" if you have no knowledge or opinion about the item. Place a checkmark in the box that reflects your beliefs.

"A"	"B"	"C"	"D"	"F"	NA	Sample of Activities
						Online Easter Pageant led by the Youth Department
						Employment Ministry Online
						Men's Prayer Line
						Praise Team Broadcasts through Social Media
						Pastor's Word Videos on Social Media
						Prayer Calls (online)
						Sermon re-broadcasts via Social Media (
						Thursday Night Bible Study Live Stream Broadcasts via Social Media

Section 2 Meeting the Needs of the Church Community

For this part of the questionnaire, consider the following question - To what extent has the church met the needs of the congregation and its community over the past five years?

Instructions: Answer the following questions about the church. Rate each item using the following scale: 5=Strongly agree (SA), 4=Agree (A), 3=Somewhat agree (SM), 2=Disagree (D), 1=Strongly disagree (SD). Place a checkmark in the column that corresponds to the way you feel.

Place a checkmark in the box that reflects your beliefs.

#	Item	SD	D	SM	A	SA
1	The cyber ministry has a positive impact on church vitality.					
2	The cyber ministry prioritizes helping people to "endure" in hard times.					
3	The cyber ministry engages in evangelism beyond its walls.					
4	The cyber ministry uses creative ways to spread the Gospel message.					
5	Online Bible study broadcasts serve an important function.					
6	Online Christian Education offerings include topics that interest me.					
7	The Sunday worship services fill a need I have.					
8	I feel that I am part of a cyber ministry "family."					
9	The cyber ministry offers programs that help meet my spiritual needs.					
10	The cyber ministry meets my expectations of what a cyber ministry should be.					

Cyber Outreach

Section 3 Demographic Data **Instructions:** Please place a checkmark beside the appropriate answer.

1. Gender
_____ Female   _____ Male

2. Age range
_____ Under 18
_____ 18-29
_____ 30-39
_____ 40-49
_____ 50-64
_____ 65+

3. Are you a member of the Church?
____ Yes   ____ No

4. If you are a member, how long have you been a member?
____ 1 year
____ 2 years
____ 3 years
____ 4 years
____ 5 years
____ Not applicable

5. What is your marital status? (Check the appropriate response.)
____ Married
____ Separated or divorced or widowed
____ Single, never married
____ Other _____

6.   What is your racial/ethnic identity? (Check the appropriate response.)
____ African American
____ Latino/a American
____ Caucasian American
____ Asian American
____ Other

7.   Do you tithe? (Check the appropriate response.)
____ Yes   ____ No

8.   How often do you attend cyber ministry activities? (Check the appropriate response.)
____ Once a week
____ Twice a month
____ Three times a month
____ Other _____

9.   How often do you attend 11:00 worship service online? (Check the appropriate response.)
____ Once a week
____ Twice a month
____ Three times a month
____ Other _____

10. How often do you watch Bible Study online? (Check the appropriate response.)
____ Once a week
____ Twice a month
____ Three times a month
____ Never
____ Other _____

Cyber Outreach

11. How often do you review job listings in the online Employment ministry? (Check the appropriate response.)
____ Once a week
____ Twice a month
____ Three times a month
____ Never
____ Other _____

12. In what community do you live? _____

Thank you for your participation!

About the Authors

About James R. Reed III

James R. Reed III is Director of Technology and Chief Information Officer for Rivertree Christian Chapel. Jim has a unique background in computer science. He worked almost 30 years with mainframe computer systems to include analyzing business problems using system analysis, testing and developing large mainframe systems and managing information system programs in mission-critical criminal justice related applications. He was trained at some of the leading computer science programs in this country. In 2002, Jim retired from a 26-year career with the Illinois State Police, where he served in various capacities, including Bureau Chief. That same year the Agency honored Jim with the Meritorious Service Award for his work with the Federal Bureau of Investigation as a project consultant with the NCIC 2000 project.

Jim possesses a bachelor's degree in Administration of Justice with a minor in Computer Science and a master's degree in Social Justice Professions Administration. Also, Jim has been blessed with many gifts and talents, which include visual art, gourmet cooking, modeling, acting, aerobics instructing, Pilates instructing, and personal fitness training, in addition to building and fine-tuning computer systems.

Now in his second career, Jim has made a commitment to use his considerable talents in service to the Lord. He is a deacon at his local church and serves as the church's Media Coordinator. His current technological focus is on providing social media marketing for churches in the Chicago Metropolitan Area. He is passionate about his work and continually seeks ways to apply his skills to further the Kingdom of God.

About Lorrie C. Reed

Lorrie C. Reed, M. Div., Ph.D., is Executive Director of the School for Ministry. Dr. Reed is also Founder and Chaplain of the Rivertree Christian Chapel. She is the former the Associate Pastor for Christian Education at a UCC congregation in the Chicago metropolitan area. Her current ministerial focus is on faith, hope, and love — uplifting spiritual, social, educational and community mechanisms for reinforcing justice and inter-generational resilience among all people. She continually searches for linkages between Jesus and Justice in a troubled world and tries to share these insights with others in need. Her non-ministerial professional experiences include service as a secondary school teacher, high school associate principal, middle school principal, curriculum director, university professor, interim dean, and consultant.

Dr. Reed is the author of several books and holds a Ph.D. in Research Methodology, a master's degree in Educational Administration, a master's degree in Theological Studies, a Master of Divinity degree, and a bachelor's degree in English education.

She is a wife, mother, grandmother, and great-grandmother. Spending time with family is one of her favorite pastimes. She lives in Chicago with Jim, her husband of five decades. She is actively involved in developing and sharing her faith through Christian educational materials and publishing. Most of all, she finds joy in serving the needs of God's people.

[1] The Barna Group. "State of the church" (July 8, 2020). https://www.barna.com/research/new-sunday-morning-part-2/

[2] The Barna Group. "State of the church" (July 8, 2020). https://www.barna.com/research/new-sunday-morning-part-2/

[3] Virginia Villa, "Most states have religious exemptions to COVID-19 social distancing rules" (April 27, 2020). https://www.pewresearch.org/fact-tank/2020/04/27/most-states-have-religious-exemptions-to-covid-19-social-distancing-rules/

[4] The Barna Group, "State of the church."

[5] Elizabeth Dias, "A Sunday Without Church: In Crisis, a Nation Asks, 'What Is Community?'" (March 15, 2020). https://www.nytimes.com/2020/03/15/us/churches-coronavirus-services.html

[6] Pew Forum, *U.S. Public Becoming Less Religious* (2015). http://www.pewforum.org/2015/11/03/u-s-public-becoming-less-religious/

[7] Pew Forum, 2015.

[8] Pew Forum, 2015.

[9] T. Rainer, *10 Church Statistics You Need to Know* (2018). https://reachrightstudios.com/10-church-statistics-need-know-2018/

[10] A. R. Neal, Enhancing the spiritual relationship: the impact of virtual worship on the real-world church experience. *Heidelberg Journal of Religions On The Internet 3.1* (2008).

[11] C. R. Wilson, Sustaining the new small church. *Anglican Theological Review, 78*(4), 549.

[12] Wilson, 1996.

[13] Pew Research Forum, 2015.

[14] Presbyterian Church (PCUSA), *Turn Mourning into Dancing! A Policy Statement on Healing Domestic Violence and Study Guide,*

Advisory Committee on Social Witness Policy of the General Assembly Council (Louisville: Office of the General Assembly, 2001), 16.

[15] Heidi A. Campbell, "6 Traits People Value in Online Faith Communities" (June 17, 2020). https://www.churchleadership.com/leading-ideas/6-traits-people-value-in-online-faith-communities/

[16] Heidi A. Campbell, "6 Traits." https://www.churchleadership.com/leading-ideas/6-traits-people-value-in-online-faith-communities/

[17] J. Alexis, *Let's Talk Modern Evangelism in the Seventh-day Adventist Church* (May 7, 2018). https://www.sdadata.org/digital-evangelism-blog/lets-talk-modern-evangelism-in-the-seventh-day-adventist-church

[18] A. R. Neal, Enhancing the spiritual relationship: the impact of virtual worship on the real-world church experience, *Heidelberg Journal of Religions On The Internet 3.1* (2008).

[19] S. Ahn, *Digital discipleship: Ministering to a digital world* (July 26, 2016). https://www.seedbed.com/digital-discipleship-ministering-to-a-digital-world/

[20] Tech Terms Computer Dictionary (2019). https://www.google.com/search?client=firefox-b-1-d&q=tech+terms

[21] Sawaram Suthar, *The basic advantage and disadvantage of Facebook ads. The Next Scoop* (2015). https://thenextscoop.com/basic-advantage-disadvantage-facebook-ads/

[22] Margaret Rouse, *Twitter* (2019). https://whatis.techtarget.com/definition/Twitter

[23] Tech Terms, 2019.

[24] Elizabeth Muckensturm, *Pros and cons of Twitter's Periscope app* (2015). https://enveritasgroup.com/campfire/pros-and-cons-of-twitters-periscope-app/

[25] Statisica, *Access over 1 million statistics and facts* (2019). https://www.statista.com/

26 S. Gelfgren, Virtual Churches, Participatory Culture, and Secularization. *Journal of Technology, Theology, and Religion*, 2(January 2011), 1-30.
27 J. L. Patterson, S. C. Purkey, and J. V. Parker, *Understanding and assessing the culture of the organization. Productive school systems for a nonrational world* (Alexandria, VA: ASCD, 1986).
28 L. C. Reed, *How big is her God? A constructive theology* (Oklahoma City, OK: Allen Carey Associates, LLC, 2014).
29 Reed, 2014.
30 Terrence E. Deal, Reframing Reform, *Educational Leadership 47, no. 8* (May 1990), pp. 6-12.
31 Terrence Deal, 1990.
32 Terrence Deal, 1990, p. 9
33 Thomas J. Sergiovanni, Why we should seek substitutes for leadership, *Educational Leadership, 49*(5), 41-45.
34 F. Newport, *Church leader and declining religious service attendance*. Polling Matters (2018, September 7). https://news.gallup.com/opinion/polling-matters/242015/church-leaders-declining-religious-service-attendance.aspx
35 Robert K. Greenleaf, "What is Servant Leadership?" (2007). https://www.greenleaf.org/what-is-servant-leadership/
36 R. K. Greenleaf, *The servant as leader* (Atlanta, GA: Greenleaf Center for Servant Leadership, 2007).
37 Greenleaf, "Servant Leadership."
38 Greenleaf, "Servant Leadership."
39 Greenleaf, "Servant Leadership."
40 Greenleaf, "Servant Leadership."
41 Greenleaf, "Servant Leadership."
42 Greenleaf, "Servant Leadership."
43 Greenleaf, "Servant Leadership."
44 Greenleaf, "Servant Leadership."
45 Greenleaf, "Servant Leadership."
46 Greenleaf, "Servant Leadership."
47 Greenleaf, "Servant Leadership."

[48] Greenleaf, "Servant Leadership."
[49] J. W. Creswell, *Qualitative inquiry & research design: Choosing among five approaches* (3rd ed.) (Los Angeles, CA: SAGE., 2013).
[50] M. Q. Patton, *Qualitative research & evaluation method* (3rd ed.) (Thousand, CA: SAGE, 2002).
[51] N. V. Ivankova, *Mixed methods applications in action research: From methods to community action* (Thousand Oaks, CA: SAGE, 2015).
[52] J. G. Brook and M. G. Brooks, *The case for constructivist classrooms* (Alexandria, VA: ASCD, 1993).
[53] C. B. Myers and L. K. Myers, *The professional educator: A new introduction to teaching and schools* (Belmont, CA: Wadsworth Publishing, 1995).
[54] J. Mezirow, *Transformative dimensions of adult learning* (Kindle Edition) (San Francisco: Jossey-Bass, 1991, 2009).
[55] Thomas Kuhn, *The structure of scientific revolutions* (2nd enlarged ed.) (Chicago: University of Chicago Press, 1970).

Manufactured by Amazon.ca
Bolton, ON